"Donald provides a treasure trove of information applicable to students across the academic spectrum. For example, he presents the rationale for using logic, not rote memorization, to solve problems, and he discusses the importance of getting the most from a textbook. Students will find the information in this book invaluable!"
—*Professor Saundra McGuire, author of* Teach Yourself How to Learn

"Quantitative problem-solving skills are essential for success in introductory science courses. Prof. Donald's text offers a helpful guide for first year undergraduate students on the necessary basic mathematics and general strategies, as well as explaining how students can more effectively study and communicate their scientific results."
—*Professor Joshua Schrier, Fordham University, New York*

How to Solve a Problem

This concise and accessible resource offers new college students, especially those in science degree programs, guidance on engaging successfully with the classroom experience and skillfully tackling technical or scientific questions. The author provides insights on identifying, from the outset, individual markers for what success in college will look like for students, how to think about the engagement with professors as a partnership, and how to function effectively in that partnership toward achieving their pre-defined goals or markers of success. It is an ideal companion for science degree prospects and first-generation students seeking insight into the college experience.

- Offers transferable problem-solving ideas and skills applicable for other disciplines and future careers.
- Provides new students with support and inspiration for their college experience.
- Includes guidance for successful interactions with professors, peers, professionals, and others.
- Encourages thoughtful determination of desired outcomes from the college experience and shaping one's actions toward accomplishing those objectives.

How to Solve a Problem

Insights for Critical Thinking, Problem-Solving, and Success in College

Kelling J. Donald

CRC Press
Taylor & Francis Group
Boca Raton London New York

CRC Press is an imprint of the
Taylor & Francis Group, an **informa** business

Designed cover image: © Shutterstock

First edition published 2023
by CRC Press
6000 Broken Sound Parkway NW, Suite 300, Boca Raton, FL 33487–2742

and by CRC Press
4 Park Square, Milton Park, Abingdon, Oxon, OX14 4RN

CRC Press is an imprint of Taylor & Francis Group, LLC

Library of Congress Cataloging-in-Publication Data
Names: Donald, Kelling J., author.
Title: How to solve a problem : insights for critical thinking, problem-solving, and success in college / Kelling J. Donald.
Description: Boca Raton : CRC Press 2023. | Includes bibliographical references and index.
Identifiers: LCCN 2022046403 (print) | LCCN 2022046404 (ebook) | ISBN 9781032203614 (paperback) | ISBN 9781032203683 (hardback) | ISBN 9781003263340 (ebook)
Subjects: LCSH: Critical thinking—Study and teaching (Higher) | Problem solving—Study and teaching (Higher) | Science—Study and teaching (Higher) | Study skills. | College student orientation.
Classification: LCC LB2395.35 .D66 2023 (print) | LCC LB2395.35 (ebook) | DDC 370.15/2—dc23/eng/20221123
LC record available at https://lccn.loc.gov/2022046403
LC ebook record available at https://lccn.loc.gov/2022046404

ISBN: 978-1-032-20368-3 (hbk)
ISBN: 978-1-032-20361-4 (pbk)
ISBN: 978-1-003-26334-0 (ebk)

DOI: 10.1201/9781003263340

Typeset in Times New Roman
by Apex CoVantage, LLC

Dedication

---

for
My Foreparents

who, in solving problems threatening basic survival,
afford my generation the space, access, and resources
to voluntarily tackle other types of problems

and

My Teachers

Contents

Preface   xiii
Acknowledgments   xvii
Author Biography   xix

1 On Encountering a Problem   1

What Is a Problem?   1
The Right to Propose a Problem   2
    The Implicit Faculty Commitment   3
    The Responsibility of the Problem Solver   4
        The Student's Personal Commitment   4
Preparing for Problems   5
Patience, Persistence, and Problem-Solving   6
Knowing $\leq$ (The Battle)/2   7
To Take on a Problem   9
    What's Really a Blessing?   13

2 The Logic of the Problem: Good Thinking and Its Rewards   19

Subject-Independent Logic (Subject-Specific Laws)   19
Scientific Laws 'Do' Nothing   21
General Logical Ideas in Science   22
    The Logic of Limiting Factors   22
    The Logic of Amounts   23

Units – The Basics   24
Units and Meaning   25
Logic above Memorization   27
Reading a Chemical Formula – Not Only
for Chemists   29

**3 Solutions in Words: Answering Short
   Answer Questions**   **33**

Symbols and Words   33
Short Answers in Words   35

4 Making Textbooks Pay   **39**

**5 Solutions in Numbers: Basic Mathematical
   Procedures**   **45**

Some Mathematical Reminders   46
    (1) Algebraic Manipulations and Some Useful
    Math Relations   46
    (2) Trigonometric Ideas   47
        Beyond Triangles   48
    (3) Other Interesting Relationships
    and Definitions   53
        Helpful Definitions and Quantities   54
More Emphasis on Logarithms and Powers   56
Linear (Straight-Line) Equations   56
Quadratic Equations   58
Graphical Representations of Experimental Data   61
Simultaneous Equations   62
    Option 1 – The Exponential Form: $A = A_o e^{-kt}$ 63
    Option 2 – The Straight Line Form:
    $\ln(A) = \ln(A_o) - kt$ 63
        An Extra Example   64
A Word on Matrices   65
On the Shapes of Things   67
    Circles, Cylinders, and Spheres   67
    Triangles and (Triangular) Prisms   68
    Rectangles and Cuboids   68
Layer upon Layer   69
    A Fun Illustration from Shapes   70
    Stay the Course   71

6 Practical Solutions: Science in the Laboratory   75

Why Experiments Matter   75
Approaching Laboratory Activities   79
   Insist on High Standards of Logic
   and Reasoning   80
   Be Willing to Think Independently
   and Take on New Challenges   81
   An Appreciation of Errors   81
   Another Suggestion to Keep in Mind   83
   The Unknown Possibilities   83
   Ethical Engagement   84

7 Spreading the Word   87

Preparing Papers   88
Writing Abstracts   89
Preparing Posters   91
Preparing Talks   95

8 Persisting against Problems   101

Mindset and Anxiety about Belonging   101
Thoughts on Managing the Demands   103
   Avoid Overcommitment   103
   Get a Calendar   103
   Sleep – Eat Well – Exercise   103
   Nurture Good Friendships   103
   Be Good to Others   104
   Remember Why You Are in College   104
   Be Gracious and Forgiving   105
   Celebrate   105
On to the Next Problem   105

*Appendix I   Additional Notes on Matrices
         and Matrix Algebra*   107
Appendix II   Thinking about Vectors: Basic Notes   111
Appendix III Safe Problem-Solving   123
Index   127

Preface

If a student makes it into a freshman college classroom, the assumption is that the student can succeed academically. As each course gets underway, it falls to the instructor to teach well and to support the student in learning well. It falls to the institution to provide opportunities and resources, and to create a context that encourages and facilitates success, but it falls to those in the arena – the instructor and the student – to make use of those opportunities and resources, in line with our respective roles, to achieve desirable academic outcomes.

For professors in that instructor-student partnership, one of the perennial questions is how to provide students with the relevant support that they need to prosper in a given course or in their work on a particular project. One reason that the question is always under consideration is that there is no single eternal answer, no universal salve that works equally well all the time for every single course. The best approach that an instructor can take, therefore, is to consider each context and identify, based on the pedagogical literature, experience, and insight, relevant and practical strategies to help students succeed.

And what of the student's position in that partnership? Each course and each professor is different, and students work out quite quickly that different approaches are needed in different cases to achieve successful outcomes. Courses may have, for example, different structures, unique stipulations (for types of assignments or modes of completing and submitting them), or even different ways to gain or improve grades (if attendance or participation are mandated or not), and so on. Then there is the actual work of learning and performing in the course itself. So, students too must consider how best to position themselves to succeed in each course. And, thankfully, students will generally do all of that and more, if they feel and in fact are supported in that effort.

These notes address particular areas of the academic demand on students in the college classroom: problem-solving, critical thinking, and general aspects of generating and reporting scientific results. After over a decade of teaching, advising, and mentoring undergraduates in traditional courses and scientific research, I found that I had accumulated pieces of notes here and there from interactions with students on how to think about and tackle one chemistry question or another. These came sometimes from informal conversations on topics in introductory or physical chemistry, discussions during laboratory sessions, office hour help sessions for problem sets, or reviews of practice exams. Thinking about some of those encounters, a few general themes seemed to emerge, and I thought it would be helpful to assemble those notes and any helpful perspective or strategy that I have considered or shared over the years all in one place where students could encounter and revisit them as needed. In many of those interactions, for example, I noticed that it was often a single missing piece of insight that caused a problem to seem intractable to students, a lack of confidence in pushing a problem to its logical end (yielding sometimes at the penultimate step), or the need for a more orderly and systematic approach to solving problems. Being unprepared in the regular sense – not attending classes, not reading notes or textbooks – was typically not the reason for the problem-solving difficulties mentioned above. This book is a response to those observations and to many mentoring conversations that I have had with students in teaching and research settings.

Students from myriad high school backgrounds converge in our increasingly diverse college classrooms. They are all called on to perform at high levels academically from the outset, and some are inevitably more prepared than others based on the quality of their high school experiences. The hope is that these notes will be useful in supporting students across that spectrum, including students who feel underprepared, to orient themselves to ways of thinking about and encountering, with confidence, the culture and academic demands of the college environment.

Each chapter in the book may be read independently, though some basic ideas do carry over from one chapter to the next. We sometimes develop mental barriers to written technical

questions based on how we perceive them (as too long, or having a lot of jargon or symbols, etc.) before we settle in to read
or start to strategize to solve them. In the earliest chapters, the
book offers some suggestions for making the most of the information provided directly and indirectly by a problem itself. The
reliance on the thread of logic in a question, even in cases where
the full path to a solution is not immediately clear, is encouraged, and hints are provided for handling questions that require
transparent explanations in prose or 'short answer' form rather
than in a stream of equations. Textbooks are severely under-
utilized if we consult them only when the professor mandates
certain chapter or page numbers, and 'Making Textbooks
Pay' encourages students to reconsider what textbooks offer –
whether they are free electronic versions or costly tomes.

Questions that rely heavily on mathematical skills can be
major sources of problem-solving challenges for students in
the introductory college science classroom. Memorization and
studying to the test, which may have been reliable strategies for
some high school classes, are approaches that will rarely work
to achieve the highest levels of success in college courses. In
such cases, the imperative will be to understand core principles
and concepts and to apply them in solving a variety of problems. A relatively long section is dedicated to solving mathematical problems. It weaves problem-solving strategies into
a conversation on critical thinking that runs throughout the
book, and the chapter integrates an overview of key topics in
mathematics that are likely to appear and reappear for science
students everywhere. Even for students who take mandatory
mathematical courses before launching into biology, chemistry, or physics degree programs, for example, this overview
may serve as a handy reference, offering helpful reminders of
useful mathematical concepts and insights into other skills and
practices that are vital in college. A brief affirmation of the
experimental nature of science, the various approaches to sharing the result of scientific investigations, and some additional
thoughts on problem-solving are included in later chapters.

This work is intended for students, yet it is not a textbook
and definitely not a review of the contemporary literature on
academic skills. The focus is on how to think about problems,
even as factual scientific and mathematical information is provided along the way. Ideally, it will yield, especially for students near the start of the freshman year, some perspectives
and strategies to strengthen their growing problem-solving
skills and maturing habits of mind as they move toward their
desired academic outcomes.

The reliance on rote learning over critical thinking, the programmatic plugging of values into formulas, and the mechanical application of prescribed procedures to a problem without understanding, are strategies that can lead to some success in high school classrooms and even in college. But that general approach may build ceilings for future learning and can close off pathways of thought that would allow students to apply old knowledge more readily to new problems. Yet, it's not a choice between one and the other. Memorization has its place – some things have to be remembered,[1] like the names and order of the planets. There is, after all, no real reason why the planet Mars could not have been called Bubble-Gum! But beyond a knowledge of the facts about Mars, understanding why it rotates and revolves frees the learner to think more meaningfully about other less familiar objects that rotate or revolve in similar or different, faster or slower ways, and to draw on transferable insights (from an understanding of Mars) to answer new questions.

To be sure, the extent to which understanding is accomplished beyond rote learning in the college classroom is not only a function of student interest and engagement. It is influenced strongly by the goals of the course and the teaching strategies employed, including the kinds of assessments used.[2] If students can succeed with rote memorization and verbatim regurgitation only, many will. But myriad tools have been developed to help instructors find creative and active ways to teach and construct effective assessment tools that require students to do more than echo their course notes. The goal here is to support new undergraduate students in their growth as problem solvers, especially those who enter college full of intellectual energy and skill, but with minimal insight (without training from college graduates in their families, high school mentors, or others) into the habits of mind and practices that lend themselves to success in college science.

NOTES

1 For a strident argument on the place of rote learning in chemistry see: Battino, R. On the Importance of Rote Learning *J. Chem. Ed.* **1992**, *69*, 135–137.

2 Elby, A. Another Reason that Physics Students Learn by Rote *Phys. Educ. Res. Am. J. Phys. Suppl.* **1999**, *67*, S52–S57.

Acknowledgments

The development of my own thinking on problem-solving and some of the ideas shared here have been influenced by the many undergraduates that I have taught in traditional classrooms and in research settings, and by colleagues with whom I have discussed teaching and learning (in learning communities and in hallways or other informal settings) over the years – at the University of Richmond (where I continue to teach with and learn from colleagues in the Gottwald Center for the Sciences), and before that in temporary teaching positions in the (pre-medical) foundation program at the Weill Cornell Medical College in Qatar, and at the University of the West Indies, Mona, in Jamaica.

I advise problem solvers to count their blessings. The blessings that I've received in my own intellectual formation are probably too many to count but I will mention some. My mathematics and science teachers at St. Catherine High School were crucial in shaping my early approaches to problem-solving.[1] I am grateful as well to my many excellent undergraduate instructors at the University of the West Indies, Mona, where academic interest was transformed into intellectual independence, and to my graduate and postdoctoral research mentors, from whom I learned many lessons about tackling problems – from planning and persistence to celebrating progress, and so much more.

As part of the first-generation in my family to enter college and graduate school, I am grateful to my parents and generations before them, who – by confronting and solving more existential problems – opened up new paths for us to choose to solve (against lower barriers, and with greater comfort and selectivity) different classes of problems.

Special thanks to Jim Davis, John Gupton, Dwayne Henry, Ovidiu Lipan, Saundra McGuire, Joshua Schrier, and Ziad Shafi for reading sections or full drafts of this work and for their generous feedback and helpful suggestions.

Kelling J. Donald
Richmond, VA
August 2022

NOTE

1 To the point and well-intended, if also gory, "There are many ways to skin a cat" is one of my early lessons in problem-solving from high school. I associate the quote most closely with Julyne McKenzie-Innis' physics classes, but it proved useful for other subjects and is relevant to many aspects of life. The point of that maxim: a problem can have many valid routes to a solution.

Author Biography

Kelling J. Donald is a professor of chemistry, currently Clarence E. Denoon Jr. Chair in the Natural Sciences, and Associate Dean in the School of Arts and Sciences at the University of Richmond (UR). A theoretical chemist by training, he teaches students across the undergraduate chemistry curriculum, in introductory and physical chemistry courses, and he mentors undergraduates in research, employing theoretical and computational approaches to address problems in structure, bonding, and reactivity in chemistry. Among other acknowledgments of his work with undergraduates, he has received the Distinguished Educator award from UR and the Henry Dreyfus Teacher-Scholar Award from the Camille and Henry Dreyfus Foundation. Born in Jamaica, he lives in Richmond, Virginia.

CHAPTER ONE

On Encountering a Problem

What Is a Problem?

Many different kinds of things are called problems. For this engagement, we are considering challenges that invite you to demonstrate and apply knowledge in an academic discipline. The focus here is intellectual problem-solving, be it for real-world applications or for classroom assessments of learning. We consider problems in science and mathematics primarily, but the key principles are applicable to other disciplines as well.

Problems usually present themselves by blessing you with a body of information that you are called on to fashion into a valid solution. The challenge comes when you are asked to outline coherently such a solution, especially if conditions are threatening to thwart your efforts or close your window of opportunity. That thwarting may come in the form of a time-keeper in an exam or the limits of your own patience. Scientific problems in the undergraduate classroom, however, are usually friendlier than we might think when we see them for the first time. That is because instructors typically ensure that questions on classroom activity sheets, tests, and exams come with routes to successful resolutions that are (or should be) well within the grasp of students, even if those routes are not obvious. In some cases, a student may even find valid approaches or solutions that the instructor did not quite anticipate.

Consider the following problem:

Find the value of x for which $0 = 2x^2 + 4x - 6$.

DOI: 10.1201/9781003263340-1

1

There are many possible initial responses to this problem – crying (in your heart or literally), dread, interest, or joy and elation – depending on your levels of focus, engagement, bravery, confidence, and preparedness. How does joy become an option? Well, you might notice immediately that, since $x^2 = 1$ when $x = 1$, $2x^2 + 4x - 6 = 6 - 6 = 0$ as required. So, hurray! $x = 1$ has to be an answer!

That strategy is a mature application of the rebel among solution strategies – the guess and check method. The method works by simply testing possible options and using previous tries to inform future guesses. Clearly, $x = 0$ would not work since if $x = 0$ in the equation $0 = 2x^2 + 4x - 6$ we get $0 \neq -6$, which is wrong. So, what of $x = 1$ as a possible solution? Let's see: Is $0 = (2 \times 1^2) + (4 \times 1) - 6$? Well, it is indeed $(2 \times 1^2) + (4 \times 1) - 6 = 6 - 6 = 0$.

So we have a solution, $x = 1$! In this case, therefore, a close look at the problem or a conscious application of a so-called 'trial and error' or 'guess and check' strategy[1] could lead to a solution.

There are definitely more structured approaches, such as the quadratic formula, $x = [-b \pm (b^2 - 4ac)^{0.5}] / 2a$, (where, for the equation $2x^2 + 4x - 6 = 0$, $a = 2$, $b = 4$, and $c = -6$), and that formula would yield two solutions ($x = 1$, and $x = -3$). Yet, how wonderful is it that just a close reading of the question could get you halfway there, with no memorized strategy or special formula required at all.

The issue with challenges, though, is that they are only attractive after you have developed some humble bravado and skills. And those come through preparation and practice. Preparation inspires confidence, which leads to small successes, which promotes confidence, which leads to more success, and so on up the virtuous spiral. You will meet problems that seem intractable or resist your efforts, but discipline will increase your win rates. Success is not a simple function of natural intelligence; it's realized through focused work – intentional preparation and practice.

A simple step that you could take immediately after reading the problem above (which we did without saying it) is to rewrite the equation $0 = 2x^2 + 4x - 6$ in a way that does not change it but that you might find to be more appealing, familiar, logical, or straightforward, such as $2x^2 + 4x - 6 = 0$. Reining a problem in by reorganizing or manipulating the information that you are given – well beyond just writing an equation the other way around – is one of the skills that you can strengthen with practice.

The Right to Propose a Problem

Your professors are likely to know that your trust has to be earned and is not necessarily an automatic by-product of their academic credentials. You can expect, therefore, that they have made commitments along the following lines to all of the students in their classrooms.

THE IMPLICIT FACULTY COMMITMENT

- **To lead** by example in our intellectual engagement; supporting your academic mission with integrity and the highest ethical standards in our academic instruction throughout our work in the course.
- **To serve** as a supportive advisor and mentor in this phase of your intellectual formation as a critical thinker and an agent for positive change in the world.
- **To prepare** well and show up for scheduled classes and meetings; providing a pedagogically sound, deep, and meaningful exposure to the subject toward an exceptional overall learning experience.
- **To foster an atmosphere of trust** in which your voice is welcomed, heard, considered, and interrogated respectfully, honestly, and fairly in all of our interactions.
- **To teach as promised** in the syllabus for the course; providing an accessible, yet rigorous treatment that prepares you appropriately for future engagements (toward a degree or otherwise) with the subject.
- **To provide useful and productive feedback** on your progress during the course.
- **To encourage your success** in other ways as needed by offering or directing you to useful resources, and making appropriate accommodations.
- **To transfer an appreciation for the subject**, even a love for it, as a fruit of excellent teaching, and exposure to its value in society.
- **To engender or strengthen transferable skills** and habits of mind – intellectual independence, working effectively in teams, good reading, writing, analytical, presentation, argumentation, study, and other broadly applicable skills and practices.
- **To identify and help to open up** unique windows for growth and development in line with your personal and professional goals.

There are, to be sure, any number of other commitments that professors make to you as a student in their classes. These covenant statements reflect, however, the kind of commitment that I and many other college professors will seek to fulfill in our partnership with you.

THE RESPONSIBILITY OF THE PROBLEM SOLVER

Your professors will have definite commitments to your prosperity and persistence in the undergraduate program, and you will have a strong commitment to your own success as well.

Consider writing down some of those personal commitments that you will make to yourself. Be as clear, realistic, specific, and measurable as possible. "I will be a good student" is vague and difficult to measure. "I will attend all of my classes" is clearer and more measurable.

Whether you write them down or not, it will be your job to assess, refine, and fulfill those commitments to yourself in your college experience. As you think about the kinds of commitments that you might write down, consider, perhaps, the following:

- Your desire for an excellent academic record at graduation (the roles that class attendance, disciplined study, office hour meetings, peer support, and so on can play in fulfilling that desire).
- The value of engaging fully in the college academic experience (participating in discussions, debates, and other forums, inside and outside the classroom, where there are opportunities to learn with and from others).
- The importance of balance, rest, and personal health and fitness.
- The possible benefits of other opportunities for mentorship and personal growth such as research opportunities with faculty, or attending public lectures on the college campus that intersect with your intended major, and even some that are outside of your academic focus but sound intriguing.
- The rewards of suitable co- and extra-curricular activities that enhance rather than detract from your primary college objectives.
- Your plans beyond graduation (e.g., graduate school, launching your career, or some other crucial next steps).

The Student's Personal Commitment It's understood that any goal might be missed due to exigencies, such as a fever or needing to travel home urgently. You can still express your commitments categorically, assuming normal circumstances. CAUTION: It will be important, however, to resist the temptation to reclassify elements of indiscipline – such as staying up playing computer games and missing an

early class the next morning (as distinguished from an illness) –
as a valid excuse for not meeting your goals. What are your
commitments? Make your list. However flexible or rigid you
make it, be sure that it is in line with your values and your
desired outcomes for the college experience.

- [Example] I will attend all my classes.
- [Example] I will complete assignments on time.
- [Example] I will play volleyball on weekends – just for fun.

Preparing for Problems

Even if you are up against the grandest of challenges, you
develop some eagerness to take them on if you feel that you have
prepared adequately. Spectators in stadium seats are sometimes
more anxious than athletes on the field because each athlete has
spent months or years preparing – much of which the onlookers
know nothing about. Faithful preparation increases your ability
and often your willingness to take on challenges. That does not
mean that you will feel no apprehension when you are on the spot
to answer a question in class or before an exam. And you are not
alone in those feelings – most of your professors felt the very
same anxieties when they were students. Exams and other forms
of assessment can be stressful. Yet the more effectively we prepare, the better we do. Now, '*more effective*' does not necessarily
mean *more time*. And it definitely does not mean applying more
stress to yourself or being stressed by others. It means having a
plan of action that aligns well with the kind of learner that you
are, the kind of assessment for which you are preparing (a written exam or seminar presentation, for example), and the context
in which you live and study.

There are many sources for study tips and ways to discern
the kinds of learning practices that might work for you.[2] Many
universities have academic skills centers, for example, that are
designed to help you to find practical strategies for success. In
college, a plan to simply memorize notes and regurgitate the
facts in exams is usually a plan to fail. Even if that approach
works for isolated courses, it will ultimately shrivel up your
capacity as a thinker and stifle your potential as an educated
scientist. The goal should be to know and understand, not just
to hear, see, and remember.

Work to understand the foundational principles of your
subject, see how its edifice builds on those foundations, and
allow the subject to open itself up to you. Getting a real understanding of what's going on in the earliest college courses in a

subject is crucial, therefore, since that's where the foundations are built. This means that good teaching is key at those levels, and so too is good learning. Once the foundations are solid, the bricklaying, in subsequent years, will be more straightforward. There will always be things that you just have to know and remember – like the names of chemical elements, for example. A chemist simply has to know that Mg is magnesium and Mn is manganese. No way around that. But the ability to complete and balance the following acid-base chemical equation is a very different thing.

$$Mg(OH)_2 + H_3PO_4 \rightarrow \underline{\hspace{1.5cm}} + \underline{\hspace{1.5cm}}$$

Your goal should be to understand why we would put $Mg_3(PO_4)_2$ and H_2O on the right-hand side and write the balanced chemical equation as follows (though, at this point, without other notations that chemists tend to include, like state symbols):

$$3Mg(OH)_2 + 2H_3PO_4 \rightarrow Mg_3(PO_4)_2 + 6H_2O$$

How a chemist arrives at this final result should be *understood*; there would be no point in simply *memorizing* the answer.

Patience, Persistence, and Problem-Solving

Nobody wants to spend time trying to solve an unsolvable problem, but you are unlikely to meet many such problems in the course of your undergraduate education. A math major might be asked to show that no solution exists for a certain equation, but the typical problem that you will face on assignments, exams, and so on, even the most challenging, will have accessible solutions for the level of the course, so do not give up on a problem just because an answer is not shouting out to you from the page! Do not yield because the solution does not spring forth after a cursory glance at the question! As you confront a question, rely on your general intellect and preparation. Position yourself as the authority, and – if one of you needs to be – allow the question to be intimidated. That positive posture requires you to assume an ability to solve the problem, it promotes confidence and allows you to conserve some of the mental energy that's usually sacrificed to anxiety.[3] Facing 30 multiple-choice questions, for example, it may be better to see them as 30 stalks of flowers to be carefully picked than thorns waiting to hurt you. That mindset is fortified, however, with effective preparation. Before most exams, the feeling will be somewhere between fear and cheer, and preparation moves you closer to cheer. You will find sometimes that, even for more difficult questions, if you press into them, joy rises as things become clearer and a solution comes into view. Such so-called 'light-bulb moments' can come suddenly – after some wrangling, you flip the switch again for the nth time and suddenly the world is as bright as can be – you have the answer. Do not nurture the habit of giving up. You may have to step away from a question from time to time, but do not do so before you have honored the problem with a serious and committed effort.

You cannot actually intimidate a problem – the ink on the paper will not quiver at your presence – but prepare, relax, strategize, and deliver. Problems doubtless will resist; keep your wits, and still persist.

Knowing ≤ (The Battle)/2

To solve any problem, you have to know something. But knowing facts alone will not be enough; in some cases, knowledge alone will be even less than half the battle.

Knowledge is definitely important for problem-solving. After all, even a distinguished chemistry professor, if she does not read German, will not be able to confidently answer this basic question: *Ist Stickstoff ein Element?* In English, the question is simply, "Is nitrogen an element?" But the great chemist would not know that. Yet, once the question is translated, even a struggling English-speaking chemistry student would reply with an enthusiastic "Yes!" So, what made the difference? Knowledge of four German words! Indeed, even if our erudite professor knew the meanings of *Ist, ein*, and *Element*, the lone word *Stickstoff* would be enough to unsettle her. One simply has to know the meaning of that word to be sure about the answer, regardless of one's many degrees in and ability to master the subject.

Although the sciences (and other academic disciplines for that matter) are not languages in the sense of English or German, they employ, in any given language, certain specialized words that are rarely used outside the disciplines (like *antigen* in biology, or *olefin* in chemistry), and some common words are often given special or technical meanings that students simply have to learn and know (like *moment* in physics, or *complex* in psychology). This is what people refer to when they talk about the 'language of a subject.' Indeed, most spheres of activity – carpentry, law, sports, etc. – develop such 'languages' that only their insiders know. If you have ever tried to explain cricket or American football to someone who has never watched the sport you know the barriers that specialized terms (the insider language) can present even for people who both speak English. The word 'frequency,' for instance, might bring very different ideas to the minds of physicists and statisticians.

Thus, the earliest courses in many disciplines will usually expose students to exciting ideas in the field while also introducing them to the language of that discipline. So, again, knowing is definitely part of the battle, and we learn the language by exposure to the subject and practice. But knowledge must often be weaved into solutions to problems by clear, creative, and systematic thinking, and that – applying critical thinking and logical analysis – is the rest, and often the larger part, of the battle. Consider, for example, the following directive:

Propose four distinct isomers with the chemical formula C_4H_8.

No German here, but if you only have facts about C and H as elements, that will not be enough. If you are meeting this question for the first time, you have to apply those pieces of knowledge about C and H, and think, synthesizing the facts that you know into an answer to the question.

If you are very familiar with the language of chemistry, (i) you know what isomers are,[4] (ii) that C_4H_8 refers to compounds with 4 carbon and 8 hydrogen atoms (see Figure 1.1), and (iii) that C atoms can form as many as four bonds to neighboring atoms (with multiple bonds allowed between atoms in certain cases) but each H atom will only form a single bond to any C atom. What the question wants to assess, however, is how well you understand and can employ and apply those facts. And that ability – to understand and show understanding through application – is critical. In

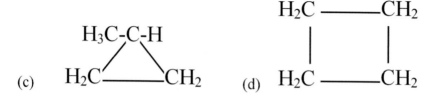

FIGURE 1.1 *Examples of isomers: Four structural isomers of C_4H_8.* These four compounds are said to be isomers of each other because they have the same number of atoms of each element (i.e. they have the same chemical formula, C_4H_8) but different arrangements of those atoms.

science, that ability is usually honed by practicing problems (on your own, with friends, for assignments, or in class, and so on). The goal is never to memorize answers, of course, but to learn the language and to understand and apply principles. In this regard, an appropriate and well-written textbook is usually an exceptional friend, since it allows you to both (i) learn the language and facts (by reading) and (ii) sharpen or assess your skills (through practice activities and sample problems).

NOTE: To help with understanding early on, before tests or exams, reading the textbook on your own, especially *before class*, is a good way to support language learning and skill-building. I emphasize *before class* here because preparing for classes gives you a chance to check with the teacher during the class on a word or concept that remains unclear despite your pre-class reading. More on textbooks later.

To Take on a Problem

In this section, I offer three pieces of advice for your consideration on the process of solving problems.

- **Study Strategies**

An important part of your formal preparation for problem-solving is studying, which is not the same as simply reading. *Studying* is an active assembling and organizing of information and working to make sense of that information in a way that enables you to articulate independently a clear understanding of the facts and to respond meaningfully to questions on the entailments of those facts. Whether the information exists as notes from classes or textbooks, public statistics on a political issue, or as written details of court cases, the same basic idea applies.

How people study varies significantly. Some of us like to have the facts spread out before us and to have time alone to analyze those facts; some prefer to coordinate with friends in the whole effort; and others tend to read alone before meeting with peers for comprehensive reviews, topical discussions, or mock interrogations (testing each other with practice questions, for example) to assess understanding and to see if they are all ready, as it were, for prime time.

An important first step in the studying process is to make sure that you have the full body of information that you need to cover. If you attend classes sporadically and do not do the work to get the information that you missed (by reading the relevant sections in the textbook, checking in with friends, or conferring with the professor on what was covered), you will be starting out with a deficit. You have automatically diminished your capacity to take on problems that would otherwise be within reach. Even before you show up for the test or make the first mistake on a question, you might already have failed it, if the test covers 100% of the material evenly, but you only studied 50% of the content. In many cases, a test does not ask questions from every single possible topic covered in the course, so your 50% might overlap perfectly with the part of the material that the test covers (Lucky you!) or it might cover only the 50% you did not study (Too bad!). The goal then should be to achieve mastery of the material (despite time constraints, competing demands, and so on), and that is best achieved by starting early with good habits such as preparing well for each class and reviewing your notes as soon as possible thereafter.

Assemble, organize, and actively study as the information comes in from a course or data collection project. You can get study tips and even detailed guidance on studying from many places, including academic skills centers and free online resources. Some of the hints that are often shared with students in college course syllabi and study skills discussions are the following.

- Read the relevant sections of the textbook or handouts before each class.
- Courses may move faster or slower than anticipated in the calendar date-to-topic schedule in a syllabus, but the sequence of the topics will usually remain the same. So as long as you know the latest topic covered in the course, you can tell from a well-structured syllabus what the next topic or objective will be.
- Do not be afraid to talk with your professor. Office hours exist for precisely that purpose. Check in with your instructor if you have a specific question, a concept from the course to discuss, a problem to resolve, or even a word of concern or appreciation for how the course is going! Your peers will want to benefit from office hours as well, and you can acknowledge and honor that by showing up well-prepared and with clear objectives for the visit; but be sure to use as needed the various opportunities that you will have to engage with your professor (in office hours, at the end of a class, etc.) during the course.
- If you miss a class, get the notes from a trusted source, or watch the recording, if that is an option. Do not expect the professor to reteach the course to you during an office hour visit, but feel free to review what was covered in the class, assemble questions on technical aspects that are still unclear, and set up an

appointment to meet with the professor during office hours to discuss those questions.

- Find an effective note-taking method (your high school strategy might not actually be the best). The Cornell note-taking method is one option to consider.[5] Others exist, with various guides available online.
- Form a study group if it works for you. Get a few friends together – three or four might be ideal – who are serious about the subject, who are willing to work independently, and who (exceptionally sharp or not) will focus when you meet to work.

CAUTION: You and your team members should make each other better students. Set clear ground rules from the start. If the group meets, but wastes time, acknowledge that, fix it, if possible, or find another group. No one profits from a study group if no learning happens there. Go to your group meetings prepared. The group should support your learning, not be your primary instructor.

Meet with the group at a practical rate: too often and you do not have any substantial new questions, too infrequently and it becomes too overwhelming to cover all of the material together. In some cases, people work alone typically and meet only for review in preparation for major tests. That works if people really have been keeping up independently with the course. Otherwise, one person will be lost at sea while watching the others paddle happily to the shore (where an A on the exam awaits).

- Bring burning questions from your pre-class readings to class with you. Ask your questions as needed if the professor gets to the relevant point in the lecture and the question is still lingering unanswered in your mind.
- Go over your notes after class. Attempt too, perhaps, a few questions from your textbook or a current problem set on topics that were covered in recent classes. Checking your textbook can help a lot if a concept or topic remains unclear after reviewing your notes.

- **Assessments: Taking Tests or Exams**

As an exam approaches, the goal is to get the highest possible score with minimal anxiety during or after the exam. If you prepared diligently and are well rested, take a deep refreshing breath, smile, and take it on! As you start the test, you have a few basic options. You can begin at the beginning and work forward through it, or you can hop around and answer the easiest or hardest questions first. Many tests flow from front to back in the logical format of the course; the questions appear more or less in the order in which the topics were discussed in class. In those cases, starting at the beginning helps since your mind can track with the logic of the course. Answering the easiest questions first may boost confidence. Answering the hardest questions first gets them out of the way, and you can cruise through the others. Whatever strategy you use, make sure that, for timed assessments, you gauge your time well. Three suggestions: (i) write down or diagram what you are thinking; you see solutions more quickly when it's looking back at you from the page, (ii) do not spend too much time on a single question that remains unyielding after a serious systematic effort, and (iii) count your blessings carefully, but do not rewrite questions just to have your own record of them; rewrite (parts of) the question only to the extent that doing so helps with finding a solution.

- **Count Your Blessing**

What do we mean by counting our blessings? Consider the following question:

> John has twelve red balls, Jill has two blue balls and one red one, and Jake has three green balls and two red ones. How many balls will John have left, if he gives one ball to Jill and to Jake for each non-red ball that they have?

Simple enough, right? The initial tendency that a student might have is to go back to the question to read it again so as to begin to think about how the numbers work out. That's fine, but the need to do that is mitigated if you *write down the clues* that might help with shaping a solution as you progress through the question. I call that process *counting your blessings*. Although this question is easy to solve, it has built-in frustrations – different colors, different amounts of balls per person, and so on. But, by writing down and organizing those facts as you read, you reduce the confusion or anxiety that you would feel otherwise when you read the question. Counting your blessings as the question unfolds involves three steps (**C.S.I**):

- **Collecting** relevant data from the question, including its goal.
- **Simplifying** that data, if possible, for example, by converting units.
- **Identifying** any obvious implications of the facts.

Carrying out such steps at the outset, while you are reading the question, lays the foundation for success even before you engage fully with the question.

For the question above, 'counting your blessings' might mean writing down the following in your own words:

Goal: How many red balls will John have left??
John: 12r   Jill: 2b + 1r   and   Jake: 3g + 2r.
r – red balls, b – blue balls, and g – green balls

NOTE: John gives up one ball for each 'none red'; How many left??

Once the key pieces of information are written down, it becomes clear both in your head and to your eyes that John would give two balls to Jill, who has two blue balls, and three balls to Jake since he has three greens balls. Adding those together, that is, $2r + 3r = 5r$, it is clear that John will be left with $12r - 5r = 7r$. And that analysis was made simpler by organizing the gifts or blessings that the question afforded us before diving in to finally solve the problem.

What's Really a Blessing?

Each problem that you will meet will endow you with pieces of information (blessings) that will open the problem up in some meaningful way for interrogation at least, if not immediate solution. Your job will be to identify those pieces and organize them toward a solution. They say that questions often reveal something about the questioner. Similarly, questions often reveal something about their solutions, and it falls to you to locate (and write down) those blessings (both obvious and sub-tle) as you survey the question. Consider the following cases:

Case 1: What are the solutions for the equation $0 = 2x^2 + 4x - 6$? Find the answers without using the quadratic formula. Hint: Both solutions are integers.

This question, a twist on one we saw earlier, tells us several things:

(1)   The teacher thinks it is reasonable (not unreasonable, at least) to present us with this problem (discounting the more sadistic alternative – that the professor expects all of our attempts to fail).

(2)   It says 'solutions', so we can anticipate two valid answers since a quadratic equation cannot have more than two solutions.

(3)   Given that $0 = 2x^2 + 4x - 6$, we can make some immediate simplifying observations even before solving for x.

- Dividing by 2 on both sides of $0 = 2x^2 + 4x - 6$ does not change the solutions, so we can work with $0 = x^2 + 2x - 3$ instead, which has smaller numbers than $0 = 2x^2 + 4x - 6$.

- $0 = x^2 + 2x - 3$, so $3 = x^2 + 2x$, which makes it a bit easier to see that $x = 1$ must be a solution, since $3 = 1 + 2$.

- Any other inference or rearrangement of information that might occur to you, such as $3 = x \cdot (x + 2)$, may be noted too as you prepare to find the set of solutions.

You might be able to think of other pieces of information that you can pick up from this question. Whatever you can glean from a question, as you read it and steady yourself in your chair to take it on, you should WRITE IT DOWN.

It may not always be clear what pieces of information are crucial and what pieces are irrelevant, but – take this case, for example – the more practice that you have working with quadratic equations, the easier it will be to see the gems. Whatever you can get directly from a question, as you examine and prepare to respond to it, will increase your chances of success when you actually begin to formulate your answer.

For Case 1 above, your 'counting' notes might look like this:

Basic Information: $0 = 2x^2 + 4x - 6$ $\implies$ *maximum two solutions. Both are integers*
$\implies$ *Goal: Find the two solutions!*
Simplified form: $0 = x^2 + 2x - 3$ *Or, if this helps . . .*
$3 = x^2 + 2x.$
$x = 1$ *is one (maybe or maybe not so obvious) answer.*
Answers: $x = 1$ *and* $x = ??$

Once you have 'counted your blessings' as we call it, you can move on to solving the problem fully. Having opened your mind up to the question by that pre-analysis, you can see through it a bit more clearly. For this specific question, for instance, we have (i) a simpler equation to work with ($0 = x^2 + 2x - 3$) and (ii) the pre-analysis has actually exposed one solution ($x = 1$) already. So we are in a superb position to finalize our solution for the question. We will focus for now, however, on process rather than specific answers so the rest of this solution is given in notes at the end of the chapter.[6,7]

Case 2: Given that 20 mL of an aqueous NaOH solution reacts completely with 30 mL of a 0.60 M aqueous solution of H_2SO_4, how many moles of NaOH were used up in the reaction?

In counting our blessings here – organizing the information that this question grants us – we might make the following notes.

Goal: **Number of moles of NaOH used up in the reaction??**

20 mL (0.02 L) of NaOH was used (concentration??)

30 mL of H_2SO_4 was used; Concentration is 0.60 M.

0.60 M H_2SO_4 (aq) means 0.60 moles H_2SO_4 is in 1 L (1 L = 1000 mL) of the solution.

So, 1000 mL $H_2SO_4(aq) \overset{contain}{\Rightarrow} 0.6$ moles H_2SO_4, which means

$$1\,mL \overset{contains}{\Rightarrow} 0.6/1000 \text{ moles, therefore,}$$

$$30\,mL \overset{contain}{\Rightarrow} 30 \times (0.6/1000) = 0.018 \text{ moles}$$

∴ 0.018 moles of H_2SO_4 were used up in the reaction with NaOH.

And we know that when H_2SO_4 reacts with NaOH the relevant chemical equation is

2 NaOH + $H_2SO_4 \rightarrow Na_2SO_4$ + 2H_2O

That's what your page might look like *before you really start processing the data*. Remember that these opening scribblings are primarily for harnessing what you know directly from the question and from general knowledge of the topic. They do not have to be very formal (it's fine to use a simple arrow '→' for instance rather than anything as pedantic or elaborate as '$\overset{contain}{\Rightarrow}$'), but make your scribblings as clear as you can since they might help the grader to see how and what you were thinking, even if your eventual answer is incorrect. Notice that I start out by making a note of what the key goal or target is for the overall question. Sometimes we get so engrossed in solving part of a problem that we forget to respond to a separate part of that same problem. So, making a note at the outset of all that the question is ultimately asking you to do can be helpful. Some students just underline the 'asks' in the question itself and double-check when they are finished that they have satisfied each of them. I see no problem with that.

Everything in the box above is either from the question directly or is readily inferred by a typically introductory chemistry student, assuming that the course has progressed enough for stoichiometric problems to even make sense. Other pieces of information that occur to you as you think about the question may be scribbled as well. You might recognize, for instance, that this is an acid-base reaction, or that NaOH and H_2SO_4 will react in a 2:1 ratio, and you might remember that acid-base reactions typically form a salt and water. All of that insight, and you have only started to take on the question! This might seem like a lot of work, but it is not. All of that information is either in the question or welling up in your mind, insights that the question plus your preparation have afforded you.

A big value to counting your blessings is that you can re-express things from the question in a way that makes the most sense to you. Instead of '0.60 M H_2SO_4,' for instance, you can write '0.60 moles/L H_2SO_4' if you prefer that format. You have rearticulated the information in a way that works for you and that should increase your comfort level in taking on the problem. I am not a golfer, but it seems to me that golfers do the same kind of situational assessment (of the landscape, wind, or golf club selection) before finally striking the ball. The solution to this (Case 2) problem is provided at the end of the chapter.[8]

In summary, as you face problems, whether the problem seems to be easy or particularly complicated, start out by organizing, mentally or on paper, all of the data that you have. Write down the useful information that a question gives to you and complete any simple step that can be taken with minimal effort as you read and think about the question. Doing so – that is, counting your blessings – frees you up mentally to focus on fully answering the question. Once you have your answer, check the goals again to make sure you did what the question asked you to do.

NOTES

1 The 'trial and error' strategy (the 'guess and check' method) is a rather unreliable partner in problem-solving because, for one, it can be quite time-consuming. If we start, for instance, with a guess that is far away from a valid solution, or if our guess is changed too incrementally, we can spend a lot of time guessing. We would take

a long time to locate our solution here if we went from $x = 0$ to $x = 0.1$, $x = 0.2$, etc. through to $x = 1.0$. But you will get better at applying the method as you grow in your mathematical understanding and build up your intuition for numbers. And it helps a lot if you think systematically: Once we got the bad outcome, $2x^2 + 4x - 6 = -6$, for $x = 0$, we went in search of a value for x that would get us closer to $2x^2 + 4x - 6 = 0$. If we tried $x = 2$, we would have found $2x^2 + 4x - 6 = 10$, which moves us from '–6' to '10' on the other side of 0. So $x = 1$, a number between our previous guesses of $x = 0$ and $x = 2$, is a logical next guess, and, thankfully, it works. I call the guess and check method the rebel among solution strategies because it has the unique feature of leading to potentially very different experiences for anyone who uses it on a given problem. That experience will depend on what their initial guesses are and the maturity of their intuition for numbers and mathematical problem-solving. The first number that Kecia will check is likely to be different from the first number that Kip, Katherine, or Karl would try on the same problem, especially if the problem is substantially more complicated than $2x^2 + 4x - 6 = 0$.

2   *Teach Yourself How to Learn: Strategies You Can Use to Ace Any Course at Any Level*, Saundra Y. McGuire with Stephanie McGuire, Stylus Publishing, 2018.

3   *Chapter 5. The Role of Anxiety and Motivation in Students' Maths and Science Achievement*, Rozek, C. S.; Levine, S. C.; and Beilock, S. L. in Developing Minds in the Digital Age: Towards a Science of Learning for 21st Century Education, Educational Research and Innovation, Eds. Kuhl, P. K.; Lim, S.-S.; Guerriero, S.; Damme, D. V. OECD Publishing, 2019.

4   Isomers are molecules that are made up of exactly the same numbers and types of building blocks (atoms) but with different arrangements of those building blocks. Those different arrangements can have real consequences. Different isomers can have very different properties.

5   Cornell Note-Taking Method: See https://lsc.cornell.edu/how-to-study/taking-notes/cornell-note-taking-system/ Last accessed July 29, 2022.

6   To solve $0 = x^2 + 2x - 3$ without the quadratic formula, the best option is trial and error (systematically changing the value of x until $x^2 + 2x - 3 = 0$). Fortunately, both solutions are integers in this case.

   If $x = 2$, we get $2^2 + (2 \times 2) - 3 = 5 \neq 0$,   If $x = 1$, we get $1^2 + (2 \times 1) - 3 = 3 - 3 = 0$. So we have one solution here!

   If $x = 0$, we get $0^2 + (2 \times 0) - 3 = -3 \neq 0$, and so on, for $x = -1$, and $x = -2, \ldots$

   And, if $x = -3$, we get $-3^2 + (2 \times -3) - 3 = 9 - 6 - 3 = 9 - 9 = 0$, which is our other solution!

   So, our two solutions are: $x = 1$, and $x = -3$.

7   Incidentally, mathematicians offer us some help here in searching for or confirming our solutions. They showed (as summarized in a result called the Rational Root Theorem) that for a polynomial expression of the form $0 = p_n x^n + p_{n-1} x^{n-1} \ldots + p_0$, rewritten as

$0 = x^n + (p_{n-1}/p_n)x^{n-1} \ldots + p_0/p_n$, by dividing through by the leading coefficient, p_n, such that the new leading coefficient is one, then the rational roots (*i.e.,* solutions that are rational numbers) must be factors of the constant term, p_0/p_n. In this case, $0 = x^2 + 2x - 3$, where the leading coefficient is already one, no division is needed, and we expect the solutions to be factors of the constant, -3. So, the *options* are ± 1 and ± 3, and we showed above that the solutions are indeed two of those values, $x = 1$ and $x = -3$. The theorem affords us here, therefore, a way to confirm our answers.

8   The products are Na_2SO_4 (the salt) and water (H_2O), and balancing the chemical equation for the reaction gives:

$$2\ NaOH\ (aq) + H_2SO_4\ (aq) \rightarrow Na_2SO_4\ (aq) + 2\ H_2O\ (l)$$

So, by this balanced chemical equation, 1 mole of H_2SO_4 reacts with 2 moles of NaOH. And since we already know from our analysis that 0.018 moles of H_2SO_4 was used, we will require 2×0.018 moles = *0.036 moles of NaOH*. And since this is in 0.02 L, the NaOH concentration is 0.036 moles / 0.02 L = 1.8 M, though we were not actually asked for the molarity.

CHAPTER TWO

The Logic of the Problem

Good Thinking and Its Rewards

Subject-Independent Logic (Subject-Specific Laws)

In many of your science courses, you will have to keep certain scientific facts, observations, or mathematical relationships in mind *en route* to solving problems. And several of those facts, observations, and relationships, have been shown (often through years of detailed investigation and confirmations by scientists) to be so broadly applicable and significant that they have now been codified as formal principles, rules, theorems, or laws.[1] For example:

- *The zeroth law of thermodynamics*: Put simply, if A and C are each at the same temperature as B, then A and C are both at the same temperature.
- *The law of mass conservation*:[2] In brief, the mass of materials in an isolated physical system is constant before, during, and after a chemical reaction. Chemical reactions do not create or destroy mass.

Such scientific truth claims have important roles in science because they sum up useful facts and years of study, and we can deploy them to guide us in our work as we seek to make new discoveries:

- The zeroth law allows us, for example, to confidently build and use thermometers:[3] Consider a traditional

DOI: 10.1201/9781003263340-2

mercury thermometer (mercury inside a narrow glass tube) at room temperature. If we put it in ice water, the mercury will cool and contract until it is in what we call thermal equilibrium with the glass tube of the thermometer, and if we put the thermometer in boiling water, the mercury will expand until it achieves a new equilibrium with the glass tube. But, in each case, is the mercury (inside the tube) in thermal equilibrium with the water (outside the tube)? This question is crucial since we assign values (for properties of water) to the heights of the mercury column in the tube (0°C for melting and 100°C for boiling)[3] and build thermometers on the claim that the temperature of the water is directly reflected by the height of the mercury column, even though the mercury and the water are never in contact. So, are the mercury and water at equilibrium, if both are at equilibrium with the glass tube? The zeroth law says a definite "Yes!"

- The law of mass conservation allows us to balance chemical equations, for if one atom of element A reacts with two atoms of element E, the product will be a molecule AE_2. In chemistry, therefore, by the idea of mass conservation, we expect the chemical equation for that reaction to be $A + 2E \rightarrow AE_2$, and *not* $A + 2E \rightarrow A_2E_2$, or any other variation – we will not arbitrarily gain or lose an extra A, E, or any other atom during a chemical reaction – and the mass of AE_2 will always be equal to the sum of the masses of one A and two E atoms.

These laws are only two of several such statements that you will learn or hear about as a science student. You will find that understanding the origins or at least the basic logic of these ideas, rather than just knowing them as rules, will allow you to solve problems with greater confidence and ease. You will find too that in many cases a law in one discipline is just an alternative way of expressing an idea that is framed in other terms in another discipline. The law of mass conservation may appeal to accountants and economists, for example, since the same notion – if we have $10 and lose $6, we will only have $4 left – allows them to successfully balance books and assess budgets, but we should only apply laws to new disciplines after careful thought. The zeroth law clearly parallels the axiom in mathematics that 'If A = B and B = C, then A = C' (i.e., "Things which are equal to the same thing are also equal to one another"), which is one of the 'Common Notions' in Euclid's *Elements*.[4] But that notion does not extend to the sociology of friendships: if A and C are both friends with B, it does not follow at all that A and C are friends! Life is more complicated than Euclid's notions and number theory.

The appropriate application of scientific rules and ideas is a central condition for valid scientific investigations. In the practice of academic research, which is greatly encouraged for undergraduates, the discovery of any 'rule of engagement' by which nature operates is a most thrilling achievement. The whole goal of scientific research is to ask important questions for which the answers are yet unknown and to apply the so-called scientific method[5] in some way toward finding answers. Sometimes, you will uncover more than you hoped to find. Sometimes, you will only eke out pieces of evidence confirming that you might actually be on the right track toward an answer, or not. Scientific laws are finely tuned summaries of what we know already, and they serve as guides and tools for us in our pursuit of new knowledge.

Scientific Laws 'Do' Nothing

Laws do not act. Laws can describe, explain, and even predict observations or phenomena in science, but they DO NOT *cause* those observations. Apples were falling from trees well before Newton's first law was formulated. What Newton did was to consider that apples move down toward the ground because there is a net force acting on them in that direction.[6] We can say that our best understanding of the observation that apples fall is that apples fall because a net force acts on them in the downward direction, fully in line with Newton's law. But that law is innocent. It does not *cause* apples to fall; if we found out that Newton was wrong to explain falling apples based on a net force pulling them down, the law would have a problem, but apples would still fall from trees as they always have. Apple trees do not know or care about Mr. Newton! His law is simply a statement in line with the experimental observation that things fall down. Of course, it's convenient to say that *'Apples fall because of Newton's first law.'* The intention is understood, but seeing a scientific law that we develop as active participants in nature can blind us to accumulating evidence that the law is limited or just wrong as an explanation of things and cost us years in the eventual search for a better one. Indeed, there have been cases where scientific conclusions were venerated, only to be shown in time to be more limited than we thought them to be or wrong altogether.[7] So, while laws (and equations based on such laws) can be consistent with, give reasons for, and allow for predictions about phenomena, they are claims about those phenomena, and not causes of them.

General Logical Ideas in Science

Recognizing general logical principles in problems can go a long way to help with problem-solving. A couple cases from first-year chemistry illustrate the point.

THE LOGIC OF LIMITING FACTORS

Consider the following question:

> If 200 atoms of reactant A and 500 atoms of reactant B are mixed together, and those two reactants A and B combine in the following way, $A + 4B \rightarrow AB_4$, which reactant will be used up first (i.e., which reactant will be the so-called limiting factor or limiting reagent)?

It becomes clear that if 1 A atom reacts with 4 B atoms then 100 A atoms will react with 400 B atoms, and 200 A atoms should react with 800 B atoms. So, if 200 A atoms will require 800 B atoms, and we only have 500 B, then B will limit how far the reaction can go; and we say, therefore, that B is the limiting factor (or limiting reagent) for this chemical reaction, and correspondingly, we say that A is in excess.

Notice that we say A is in excess even though we actually have numerically less A than B atoms when we start the reaction. This is because we are guided by the requirements of the process (as shown in the chemical equation above), not the actual numbers of atoms available at the start of the process.

If you thought, however, that the question was fundamentally a chemistry question, you are wrong. It was a *logic* problem that could be posed in many other ways. The exact same analysis would apply, for instance, if you worked at a car factory (if we simplify the car manufacturing process a bit for illustration as follows):

$$1\,Car\,Body + 4\,tires \rightarrow 1\,Car.$$

Clearly, if we have 200 car bodies we would need 800 tires to make 200 complete cars. If we only have 500 tires, however, we would obviously need 300 more, so the tires are limiting us. Put another way, the car bodies are in excess.

And we could carry out the same assessment by looking first at how many tires we have: A quick check will show that 500 tires will allow you to make only 125 cars (since 500/4 = 125). So, if we are fine with that, we could calculate how many car bodies will remain at the end. That number (subtracting 125 car bodies from 200 car bodies) would be 75 car bodies, since – by our balanced manufacturing equation above – each car requires only one car body. So, although we have 200 car bodies, only 125 can be used, leaving 75 wheelless frames. The wheels are limiting us! We have 500 wheels and 200 bodies, but 'wheels' is the limiting factor because we still need more of them if we are to use up all 200 car bodies.

The same logic holds regardless of the unit that we use or the objects (atoms or cars, for example) specified in the equation. In the same way that 1 'A' combines with 4 'B' to give one 'AB_4,' one dozen 'A' would combine with four dozen 'B' to give one dozen 'AB_4,' and one mole of A would combine with 4 moles of B to give 1 mole of AB_4. *And, if we start with 0.2 and 0.5 moles of A and B, respectively, we can show easily now that 0.125 moles of AB_4 can be made, with 0.075 moles of A remaining.* But now we know that the question draws on logic, not any uniquely chemical insight.

Always consider the general logic behind the problem. I have no doubt that if a student was working in an auto mechanic shop for a summer and saw two cars without wheels and only five tires, the student would wonder, 'Wait a minute, where are the other three tires?'

THE LOGIC OF AMOUNTS

If you were asked to calculate by some means how many of a certain number of rabbit offspring will have a specified genetic trait, a hint that you are on the wrong track is that your answer is -22 rabbits. A negative number is not likely to be a correct answer to that question.

If you are asked to calculate the mass of an electron and your answer is 2 kg, that result should register as a scream from the page out to you that something went wrong (especially if you knew already that the mass of the electron is the marvelously tiny 9.1094×10^{-31} kg).

The point here, from these two examples, is that a big benefit of practice is that you become more familiar with the basic nature of your subject such that you develop an intuition for what is reasonable. If you are taking a course on the American

Civil War and you select 1914 as the year of Lincoln's assassination, that's a bad sign.

Sometimes it is clear that an answer does not work logically, such as a negative number of rabbits, or if the number of tires used plus the number of tires remaining is different from the total number of tires with which you started.

Sometimes flaws in your answer will be apparent, but only if you are very familiar with the content of the course. If you have a good idea from laboratory experiments for what 18 grams of water looks like and you recall that 18 grams of water is very roughly 1 mole of water molecules, which is 6.022×10^{23} water molecules, and each water molecule contains two H atoms and an O atom, and those are made up of electrons protons, and neutrons, then clearly a single electron could not have a mass of 2 kilograms.

There is sometimes the temptation, in working on an assignment or exam, to simply find an answer quickly and move on to another question, but it helps to ask at the end of a problem if the answer that you landed on is at least logical and reasonable based on everything else that you know. At the very least, that might allow you to leave a meaningful note to the professor, "This answer makes no sense, but I'm not sure what I did wrong." Such a note will earn you perhaps no extra point on a test, but it will show that you appreciate the logic structure of the subject. And that says a lot.

Units – The Basics

Units are blessings too, even if, as a student struggling to keep track of them in your calculations, you wish they did not exist.

One impressive and useful feature of units is that there is a small handful of basic quantities with respect to which all other quantities can be defined. Each of those basic quantities has a unit, and those fundamental units are called *base units*. These are units from which all other units that you have ever seen or heard of are made and into which they can be broken down.

One example of a fundamental quantity – one that you know very well already – is length (or distance). The corresponding base unit is the meter. We can still use other units like 'yard' or 'bamboo,' but after a number of international discussions starting as early as 1875, authorities from across the world have converged on an international system of units (now officially called SI units)[8] and the meter is the agreed-upon SI unit for length. Base units are seen as fundamental units because none can be written in terms of any combination of the others. The fundamental quantities and their units and symbols are:

Quantity and common symbol		Unit and symbol	
Time	t	second	s
Mass	m	kilogram	kg
Length	l	meter	m
Temperature	T	kelvin	K
Current (electric current)	I	ampere	A
Amount of a substance	n	mole	mol
Luminous intensity	I_v	candela	cd

I have listed the quantities here following an extension on a mnemonic that I've used since high school when I was introduced to the first five of them: "Tell Mom Let Tom Come, And Lucy." Again, these quantities are important because any other quantity that you can think about – *for example,* area, volume, energy, density, or concentration can be written out explicitly in terms of their component base units: m^2, m^3, $kg \cdot m^2 \cdot s^{-2}$, $kg \cdot m^{-3}$, and $mol \cdot m^{-3}$, respectively. Knowing the base units can be very insightful. Notice, for instance, that density ($kg \cdot m^{-3}$) and concentration ($mol \cdot m^{-3}$) are similar in that they are both measures of a substance in a certain volume; the mass in one case, and the actual amount (moles) in the other, so concentration is, in a real sense, the density of the solute in a solution.

Units and Meaning

If I say to a fruit seller, "Fifteen, please," what would the fruit seller do? Probably just stare at me.

If you said, "Fifteen *mangoes*, please," you would be in business, literally.

'Mango,' in that case, is the unit. Similarly, "Three, please," might be too imprecise at a fish market, but "Three *pounds of cod*, please," makes good sense, and in that case '*pounds of cod*' is the unit. Seems trivial? Units roll out for us some invaluable clues about how to begin to think about a problem and how to proceed toward a solution.

- *Illustration 1*: Let's assume that we give the following problem to an enthusiastic high school student who knows what the units 'm,' 's,' 'km,' and 'mins' mean, but does not know how speed relates to distance and time.

Find the speed in m/s units of a huge blue truck that covers 3.6 km in two mins.

We will approach this question in the way we discussed previously.

Counting Blessings

Goal: need speed in m/s units??
The student could make the following observations:

(1) Truck is irrelevant, huge blue or otherwise!
(2) Distance: 3 km = 3.6 × 1000 m = 3,600 m.
(3) Time: 2 minutes = 2 × 60 seconds = 120 seconds.

[*Unwritten Reflection*: What is speed? I do not know, but I have values in meters and seconds and the answer should be in meters/seconds, so I will just divide and hope for the best.]

Solution (Decisive Calculations)

Speed = Distance / Time (a deduction from the units)
So, speed = 3,600 m / 120 seconds = 30 m/s! Cool!
[*Unwritten Reflection*: Hope this is right. At least it makes sense based on the units!!]

NOTE: The headings (*Counting Blessing*, and *Solutions*) are not needed, of course; they serve only to emphasize the stages that you can follow in setting up the problem. As for exclaiming 'Cool' at the end of the problem, most professors have no trouble with you celebrating small victories on a test or problem set with a sly comment here or there, such as 'Got you!,' or the popular QED from the Latin *quod erat demonstrandum* (i.e., 'what was to be demonstrated' and parodied by 'quite elegantly (or easily) done'). Do not overuse these celebratory exclamations though. Save them for truly thrilling conquests.

- *Illustration 2*: Here's another angle on the value of units, emphasizing, in this case, the usefulness of base units. We assume for this illustration that, for whatever reason, the student does not know how kinetic energy is linked to mass and momentum.

What is the kinetic energy, E_K, in joules, of a mass, m, of 10 kg with a momentum, p, of 50 kg·m·s^{-1}.

Counting Blessings

Kinetic energy, E_K, is just another form of energy. The SI unit of energy is joule, and
   In base units, 1 joule = 1 kg·m²·s⁻².
   ***mass* = 10 kg; momentum, *p,* = 50 kg·m·s⁻¹.**

[*Unwritten Reflection (or addition to working)*: The unit of energy is kg·m²·s⁻², and the only way to get 'm²·s⁻²' from mass and momentum is to square momentum. That would give $p^2 = 2500$ kg²·m²·s⁻², but now we have kg and kg² from the units of mass and p^2, respectively, and the only way to get kg from that is to divide p^2 by mass.]

Solution (Decisive Calculations)

Evidently, therefore, $E_K \propto p^2/m$ so we can write $E_K = kp^2/m$ where k must be a unitless constant, So:

$$E_K = kp^2/m = k \cdot 2500 \text{ kg}^2 \cdot \text{m}^2 \cdot \text{s}^{-2} / 10 \text{ kg} = k \cdot 250 \text{ kg} \cdot \text{m}^2 \cdot \text{s}^{-2}$$

Sorry, I'm not sure what 'k' is, and do not recall how exactly to link E_K to p and m.

Maybe the student really should know that $k = \frac{1}{2}$, that is, $E_K = p^2/(2 \cdot m)$. I am not prescribing here a way to bypass studying – I've assumed, for instance, that you know that joules in SI base units is kg²·m²·s⁻². But, in a moment of mild desperation, the observations that we make as we write down what we are given by the question and the relevant facts that we recall (counting our blessings) can sometimes jolt our memories, bringing to mind some pieces of information that would otherwise have escaped us. The '$\frac{1}{2}$' in the formula does not have a unit, but, as we showed, we could get a long way through the problem on just an analysis of the units (a form of dimensional analysis). So, watch your units! Even if you lose some credit for missing the '$\frac{1}{2}$,' the professor will be quietly impressed with the quality of the effort, if you show the thinking that led you so close to the goal.

Logic above Memorization

When you take on problems as practice or in exams, always prefer logic over simple applications of memorized procedures.[9] Consider, for example, the following question about dilution.

A chef is making a sugar solution. He dissolves 0.4 moles of sucrose in water in a one-liter (L) container such that the concentration of the solution is 0.4 moles/L (or 0.4 M). What is the concentration of a solution that is prepared by removing 0.1 L from that 1 L solution and adding more water to the 0.1 L to dilute it to a total volume of 200 mL?

Two different ways to approach this problem may be classified as 'logical' and 'procedural.' Both can give you the correct answer, but one relies primarily on understanding while the other gives a mathematical formula into which to substitute numbers.

Option 1 (Logical)	Option 2 (Procedural)
Goal: Concentration of new solution	
1L original solution contains 0.4 moles	1L original solution contains 0.4 moles
So, 0.1 L original solution contains	Formula is: $M_1V_1 = M_2V_2$. 0.4 M is M_1;
(0.4 moles / 1 L) × 0.1 = 0.04 moles	V_1 is 0.1 L $\Rightarrow$ the volume, we removed from the original solution.
And 200 mL = 0.2 L	V_2 is the total volume after that
The new 0.2 L volume of solution is made up from 0.1 L of original solution (plus added water).	0.1 L was diluted $\Rightarrow$ 200 mL. And 200 mL = 0.2 L. So, $V_2 = 0.2$ L
****	Now, we need to find M_2.
So, 0.04 moles were transferred, and that was made up with water to 0.2 L.	**** 0.4 M · 0.1 L = M_2 · 0.2 L
So, 0.04 moles is in 0.2L and, therefore, the new concentration is:	$\therefore$ M_2 = (0.4 M · 0.1 L) / 0.2 L = 0.2 M
0.04 moles/0.2 L = 0.2 M	

Logic Check: *0.2 M is less than the original (0.4 M) concentration. And making 0.1 L up to 0.2 L is expected to reduce (halve, in fact) the concentration. So, the results seem to make sense.*

If possible, always choose logical strategies over procedural methods for problem-solving. Why? Well, not all questions that you will meet will have memorizable procedures that would lead to correct answers. Certain simple equations or procedures may apply in only very special cases, so the mindless matching of *type of problem* to *type of equation* can lead to disappointments, but reasoning your way through a problem allows you to bypass such difficulties. For example, consider the equation, $M_1V_1 = M_2V_2$, that we used above for the dilution problem. Some chemistry students use it for so-called titration calculations as well, which is a rather different situation from dilution. As it turns out, the $M_1V_1 = M_2V_2$ equation is only valid for titration calculations for 1:1 reactions (i.e., where one mole of one species reacts with one mole of the other in a titration), such as $NaOH + HCl \rightarrow NaCl + H_2O$. For such a 1:1 titration, where, for example, a known volume, V_1, of a NaOH solution of unknown concentration M_1 reacts completely with a volume, V_2, of an HCl solution of known concentration, M_2, it turns out that we can use the equation $M_1V_1 = M_2V_2$ to find M_1. But that is a convenient coincidence. The $M_1V_1 = M_2V_2$ equation is NOT valid at all for titrations that are not 1:1 reactions,[10] such as the 2:1 reaction between sodium hydroxide and sulfuric acid, which follows the chemical equation: $2NaOH + H_2SO_4 \rightarrow Na_2SO_4 + 2H_2O$.

Even if your answers end up being wrong sometimes, insist always that your answers be logical.

Reading a Chemical Formula – Not Only for Chemists

Chemical formulae summarize the number of atoms of different elements that are in a compound. They are not descriptions of how atoms are bonded to each other in a compound. For example, the chemical formula for water is 'H_2O,' but that representation does not mean that two hydrogen, H, atoms are bonded to each other in the manner H-H-O. The arrangement of the atomic centers in water is H-O-H. That is, the H atoms are bonded separately to the central oxygen, O, atom. The water molecule is known to be bent and the O atom is at the vertex of the bent triatomic molecule. The same basic bonding arrangement is found for $BeCl_2$, except that Cl-Be-Cl is linear (Figure 2.1). For good measure, we include in Figure 2.1 a model of ammonia as well. Ammonia is less celebrated perhaps than water (though it is commonly used in household cleaning solutions and in manufacturing fertilizers) but it is also a simple molecule with an interesting (pyramidal) shape.

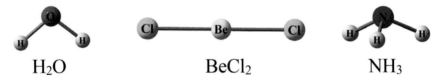

H_2O $BeCl_2$ NH_3

FIGURE 2.1 Sample structures of simple molecules.

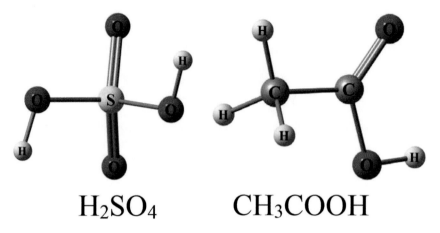

$$H_2SO_4 \qquad CH_3COOH$$

FIGURE 2.2   Sample structures of two acids.

When compounds dissociate in water or react they do not break up in the way that they are written. For example, H_2SO_4 is an acid that dissociates to release H^+ ions. It can split apart in water to give 2 H^+ ions and one SO_4^{2-} ion. It will NOT give H_2, S, and O_4. You will learn more about the structures of molecules and how compounds behave in solution or react as you progress through general chemistry. The main message here is that the chemical formula of a compound sums up the identity and number of atoms in that compound but they do not usually tell us anything explicitly about the shape, bonding, or stability of the compound. Chemical formulae can sometimes give clues, however, to structure, such as in the formula for acetic acid: CH_3COOH. This formula has a few subtle messages about structure. It indicates that one of the H atoms is different from the other three and that the two oxygens are bonded to the same C atom, but not in the same way (as we show on the right in Figure 2.2).

But one would be hard-pressed to determine the structure of H_2SO_4 (on the left in Figure 2.2) from a knowledge of the chemical formula alone.

NOTES

1 Terms like principles, rules, and laws are sometimes used loosely and interchangeably, but laws, briefly defined, are summary statements of experimental observations accepted universally as true. Theorems – you will find many of these in mathematics – are claims that have been proven based on simpler claims that are taken to be self-evidently true (i.e., axioms) or other already proven theorems.

2 We refer here to a definition in terms of chemical reactions. Nuances related to mass-energy equivalence which you might meet in physics are irrelevant for chemical reactions. We normally ignore as well, for chemical reactions, changes in the total matter in the system in the case of radioactivity.

3 Wisniak, J. The Thermometer – From The Feeling To The Instrument. *The Chem. Educ.* **2000**, *5*, 88–91.

4 *Euclid's Elements – All thirteen books complete in one volume,* The Thomas Heath Translation, Ed. D. Densmore, Green Lion Press, Santa Fe, NM, 2003. p. 2 (Common Notions – #1).

5 We say more later on about the scientific method. The term describes a formal approach used in the search for new empirical knowledge.

6 For a brief report on the story of Newton and the falling apple, see: Gefter, A. *Newton's Apple: The Real Story* New Scientist, January 18, 2010: www.newscientist.com/article/2170052-newtons-apple-the-real-story/

7 A few examples are cited in: Vickers, P. *The misleading evidence that fooled scientists for decades*, The Conversation, June 4, 2018: https://theconversation.com/the-misleading-evidence-that-fooled-scientists-for-decades-95737.

8 'International System of units' in French is 'Système International d'unités' and the abbreviation of that French rendition, 'SI,' has been accepted internationally as the official abbreviation of this unit system.

9 The power of logical thinking and the educated guesses that good reasoning skills make possible are emphasized in this book: *Street-Fighting Mathematics: The Art of Educated Guessing and Opportunistic Problem Solving*, Sanjoy Mahajan, The MIT Press: https://mitpress.mit.edu/books/street-fighting-mathematics. Last accessed July 30, 2022.

10 The $M_1V_1 = M_2V_2$ dilution equation is irrelevant for titrations that do not involve 1:1 reactions. Notice that the product molarity × volume (M_1V_1, which is [moles/volume] × volume) for a given solution returns the number of moles of the solute in that volume of solution. That number of moles of solute does not change if we dilute the solution simply by adding water, even though the volume increases of course after dilution and the concentration naturally decreases. So, it is expected that M_1V_1 (i.e., number of moles before dilution) = M_2V_2 (i.e., number of moles after dilution). For a titration, however, the numbers of moles of titrant and analyte used up in the reaction are not necessarily equal at the end of the titration. That is only the case for 1:1 reactions (such as NaOH (*aq*) + HCl (*aq*) → NaCl (*aq*) + H_2O (*l*)), where it so happens that the number of moles of one reactant used up during a titration, $M_1 × V_1$, is equal to the number of moles of the other reactant used up at that same point in the titration, i.e. $M_2 × V_2$. If the equation $M_1V_1 = M_2V_2$ is deployed for a 2:1 titration (such as 2 NaOH (*aq*) + H_2SO_4 (*aq*) → Na_2SO_4 (*aq*) + 2 H_2O (*l*)), for instance, you will get a very wrong answer.

CHAPTER THREE

Solutions in Words
Answering Short Answer Questions

Symbols and Words

In reality, all problems are word problems. Mathematical problems are word problems in which some of the words are replaced for brevity and precision by symbols. Equations are sentences in symbolic form. Consider for instance, the following:

- Format-1: Determine how many books will be left in our library if Jane donates her fifty-two books, Jamal gives us his eighty-four books, and we give Jenny the sixteen books that we promised her for her collection.
- Format 2: What is the value of y, if $y = 52 + 84 - 16$.

Both formats are asking essentially the same thing, even though we do not include the units 'books' in the second format. And we can answer the question in more or less verbose ways as well.

- Format 1: Jane's fifty-two books added to Jamal's eight-four books is a total of one hundred and thirty-six books donated. That's a lot of books, but we had promised Jenny sixteen so we're left with one hundred and twenty books for the new library.
- Format 2: $y = 52 + 84 - 16 = 136 - 16 = 120$.

DOI: 10.1201/9781003263340-3

Similarly, we can construct syllogisms in text form – saying in words that

- Format 1: All soccer players have good hand-eye coordination. James is a soccer player, so James has good hand-eye coordination.

Or, we could make the argument as follows.

- Format 2a: If $s(S)$ is the set of soccer players, J is James, and $s(C)$ is everyone with good hand-eye coordination, then where '$\subset$' means 'is a subset of,' and '$\in$' means 'belongs to,' we can write:[1]

$$s(S) \subset s(C), \text{ and } J \in s(S), \therefore J \in s(C).$$

Format 2b: If it helped us to see the relationship more clearly, we could even employ a simple Venn Diagram.

This perspective or way of thinking – that mathematics is a rendering of text in terms of symbols[2] – might help to reduce anxieties for some in engaging with equations and mathematical statements. College students rarely feel similar apprehensions when they encounter a paragraph worth of text, even if the latter is more complicated and nuanced than the equation of a straight line. A key aspect of both training and working in the sciences is to learn *over a period of time* (not all on the first day of a course) the meanings of the relevant symbols and to get used to using them – which, like so much else, is aided by

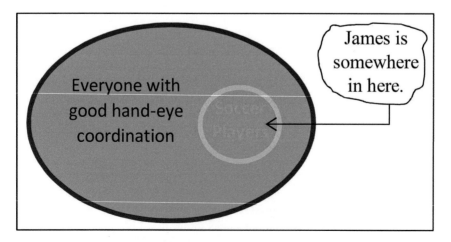

FIGURE 3.1   A simple Venn diagram illustrating the syllogism.

practice. After all, even for regular prose, we all had to familiarize ourselves early on in language learning with other symbols (punctuation marks) that we take for granted now when we read, such as those here between – and including – these parentheses (''., – ;?: – ! ''''). We are so comfortable with these symbols that we usually feel no anxiety at all when we see them in a paragraph that we must read and understand. Why are we so comfortable with them? We have spent years using them and – the great benefits of childhood education – those symbols were introduced to us before we learned to be overly anxious about symbolic reasoning and learning in general.

In mathematical representations, the use of symbols has the special advantage of making our arguments somewhat more accessible across language barriers, even if sometimes different symbols are used for numbers or certain mathematical operations across languages. Keep in mind too that this reliance on symbols is not unique to the natural sciences and mathematics. Special symbols are used to good effect to summarize information and to provide instructions in music (♯♭♩), and in any number of other areas of life.

Short Answers in Words

Even in the sciences, however, where stereotypes hold that the most fearsome mathematical equations are to be found, it is important to be able to express ideas or provide an explanation or description in words. Moreover, equations can be a bit dry. Saying the mural was breathtaking, or that soccer is a lovely game, is best left to words. Sometimes in the sciences, as well, you will be required to articulate responses with more syllables than symbols. Even in mathematics, symbolic logic and rigor are not the only virtues. A wry phrase can make even a hardened professor of pure mathematics smile.

A few tips are offered here for writing responses to short-answer questions, but please check with academic skills support professionals on your campus for reading materials and additional advice on writing well. Your broader goal in this context might be to have a high degree of confidence and a low level of anxiety concerning your ability to produce strong and well-structured (long or short) responses in words. The use of language in sharing your questions, observations, perspectives, and insights with others is a skill that is as important in the sciences as it is in other areas.

- As you would for quantitative questions, count your blessings by taking note of any insight that a so-called short-answer or 'essay-type' question provides and make sure that you are clear on what the question is asking you to do.

- As you would for quantitative questions, look for clues in the logic of the question as well.

  ⇒ *Example*: If a question starts out with, "What strategies are commonly used to . . .," one demand that it makes is that you give more than one 'strategy.' If you know of or remember only one strategy, however, be sure to state it and move on. If you feel very strongly that only one 'strategy' is *commonly* used but you are aware of the range of strategies, one enlivening response is to provide the list requested and (if you have time) state why you think the others are really not *common*. You may be rewarded for your awareness and extra analysis even if the teacher quibbles with the conclusion. Even so, you should take seriously, in general, the demands of the question as it is designed and presented.

- Confront the demands of the question as directly as possible.

  ⇒ *Example*: Consider the question, '*What is the photoelectric effect?*' Your answer to such a question might reasonably be, '*The photoelectric effect is the observation that exposure to radiation can cause electrons to be ejected from certain materials.*' You might say instead, '*The term photoelectric effect describes . . .* ' or '*This is a phenomenon in which . . .* ' You could even start by giving some background: '*At the beginning of the 1900s, people still did not know why . . . That phenomenon is called the photoelectric effect.*' You should not, however, answer this type of question in the following ways. 'It is when . . .' or 'It is like . . .,' and you definitely should not say, '*The photoelectric effect is something that Einstein explained about light and metals.*' The latter statement is true in a sense, but it is not an explanation of the effect.

- Try to avoid analogies in short definitions. Analogies can be helpful in explaining ideas, spurring and opening up the imagination, but when a short and clear scientific explanation or a definition is required, address the issue directly.

- If only a short definition is requested do not provide a long essay.
- If an essay is required, do not provide a short answer, unless that short answer is so comprehensive and so powerful that the professor will find it irresistible. Learn and use good essay writing skills. Each essay is another chance at perfecting your craft. Again, though, focus on the topic, and follow as closely as possible, in the style, structure, and length of your responses, the stipulations of the question.

How short is short? Some questions will state the expectations clearly: 'In 200 words or less . . .,' 'Explain briefly . . .,' or 'In a sentence of two, justify . . .,' and so on. Short is relative, but in undergraduate science classrooms, it is often assumed that, for a 'short explanation,' not much more than a robust paragraph is needed, unless you are told otherwise by the professor or in the question itself.

NOTES

1 The three dots in the arrangement '$\therefore$' is shorthand for 'therefore.' It is used in constructing logical arguments in various areas across the sciences and mathematics.
2 For an example of how academics have sought to capitalize on this evident parallel between equations and text see: Roy, S.; Upadhyay, S.; Roth, D. *EQUATION PARSING: Mapping Sentences to Grounded Equations* arXiv:1609.08824v1. **2016**.

CHAPTER FOUR

Making Textbooks Pay

Textbooks are usually as useful as you make them, regardless of whether they are free electronic books or expensive hard copies. The following perspectives and pointers may help you to benefit optimally from them.

- **Why use textbooks?** Right at the start of the course, textbooks provide you with an intelligent presentation of the material that you will need to learn. The information comes pre-organized – better than your notes are likely to be during the semester – in a thorough and systematic way. Good textbooks hand you the course almost on a silver platter! The book lacks the vital in-class engagement, but overly enthusiastic students have been known to master large fractions of the course material by spending the preceding summer with the textbook. Beyond such extremes, however, textbooks are great supports during the course, for both independent study and practice (if the book has practice questions), and it allows you to explore, even beyond the limits of syllabi, on your own time. One of the things that I realized early on in college was that the most academically successful upperclassmen I knew tried to make the most of their textbooks.
- **Start from the 'beginning.'** Do not consult your textbooks only for a definition here or for clarification on a topic there. A quick check to clarify a concept, or to confirm a definition, has its place for sure. If you are just being introduced to a topic, however, reading the relevant section from the textbook can go a long way to enhance your appreciation of the whole

DOI: 10.1201/9781003263340-4

subject. If you want to know what the Mpemba effect is, you could just check for a line that offers up a definition. But the concept is enriched when you read further to find the story of a curious Tanzanian kid, Mpemba, asking a teacher and ultimately a physics professor visiting his school if it's possible for hot liquids to freeze faster than cold liquids (his original question was about ice cream mixtures that he and his friends would make). When he posed his question to the visiting professor, his friends laughed at him, and his teacher was embarrassed, but he and that physics professor ended up publishing a paper together about the observation, now known as the Mpemba effect.[1] Reading more for a class than is absolutely necessary promotes understanding (which leaves less for memorization), and gives you a better handle on and greater pleasure in the subject.

Title Page   *– The title of the book and names of the authors*
Table of Contents   *– Chapter titles, insights into what each chapter has to offer, the order in which the subject is covered, and the relevant page numbers.*
Preface and **Acknowledgments**
Contents (Book Chapters) – Individual chapters of the book.   *– It is common for each chapter in science textbooks to end with practice problems specific to the material covered in that chapter, some questions drawing too on knowledge from preceding chapters.*
Appendices – Additional supporting information.   *– An appendix might provide some helpful technical notes, a more advanced treatment of a particular topic, additional evidence or supporting materials for claims made in the book, or tables of scientific constants or chemical data.*
Answers to All or Selected Chapter Problems
Glossary – Key terms and their definitions.
Index – List of words (significant topics, names, important ideas, etc.) mentioned in chapters in the book.

- **Benefit from the structure.** If you simply wanted to find the word 'oscillation' in a physics textbook, where would you look? Your first stop should probably be the index at the back of the book (see the list included in this section of basic elements of common textbooks). The index lists key words and terms that appear in the book and the relevant page numbers. Turning to the index to locate a word in the textbook is much better than going to the table of contents. Some books have a glossary at the back as well, but glossaries typically include far fewer words than indices since the goal of the glossary is to provide definitions of specialized ideas and concepts. The index does not include definitions, but it is far less discriminating. If you go to the index you might find that 'oscillation' comes up in several different chapters (on harmonic motion, waves, electricity, and electronics), and finding all of those intersections might open up for you the power and usefulness of the concept even if you had a specific use of the term in mind originally.

- **Feel free to make jottings** and to highlight or underline statements in your textbooks.[2] Unless the book was borrowed, or you wish to sell it in mint condition when the semester ends, feel free to write in it. A new textbook is more like a newspaper or a diary than a rare and delicate ancient scroll. You could decide not to attempt the crossword puzzle in the newspaper or write in your diary because you did not want to smudge the pages, but that would be sad. If a sentence in your textbook astounds you, feel free to memorialize the encounter, if you wish, by highlighting the sentence or leaving a brief mark or comment. One day, years from now, you will open that book, and you will feel a surge of anxious joy as you see a mark or a scribble, and you will be gratified to see or to read it, even if you cannot recall why you left it there. I've heard people say that they regret not highlighting a quote that they wish they could locate easily now in a massive tome that they read long ago. Few will lament a few years on that they drew a line under an impressive revelation that they met in a book. Be careful though to not equate a cursory engagement with the material – which the simple act itself of underlining or highlighting some word or section in a text can be – with learning. A shallow engagement alone can lead to an illusion of understanding[3] where you think you understand a thing, but you are, in practice, unable to explain or apply it. In that regard, incorporating practice problems as part of your study regimen can be an affirmation of understanding and an antidote to illusion. If you make side notes, feel free to make them personal and creative, if that helps you to understand and remember. An athlete might find a way to frame the citric acid cycle, for example, as a track event with different processes occurring at each stage. You might invent your own funny mnemonic for the key stages of the cycle.

If you do not want to write in your textbook, take notes separately as you read so that you can capture key ideas or solve sample questions from the book on the fly as you are reading a section in preparation, perhaps, for your next class or an approaching exam. Taking notes in class or while reading on your own is one way to guarantee that the information has traversed at least once through your mind.

- **Your textbook can be your coach.** Textbooks often have helpful tips, topic maps linking ideas in the course, extra materials such as quizzes, souped-up diagrams, charts, and so on. They often have worked examples, too, separately from practice questions for which no solution is provided.[4] To assess your mastery of a topic, feel free to attempt the worked examples first on your own before looking at the solutions provided in the book. Then you can take on the practice or end-of-chapter questions. And I suggest that you try at least two or three times to solve such questions before seeking help from anywhere else (from peers or your professors). That way, when you ask for help you can start out by showing the direction from which you (attempted to) tackle the problem. Again, go to office hours, and bring your questions.
- **Seeing might be believing, but it is not knowing.** Many of us have seen somebody do something, and – thinking to ourselves that it is easy, straightforward, or simple – failed miserably when we attempted it ourselves for the first time. Do not simply look at a solution to a problem and embrace the feeling of understanding the solution so thoroughly that you do not attempt the solution yourself. Otherwise, you will end up in front of an exam paper a few days hence, thinking, "I've

seen this before; if only I could remember how they solved it?" The mind has a way of convincing us that we know or understand things well or possess certain skills when we really do not.[5] A sure way to know that you are able is to do. Confidence without skill is not ability. Put your presumed skills to the test, and celebrate your successes. If you come up short, however, respond by reviewing the topic, practicing on simple problems, and working your way up to the required skill level at least. As gaps emerge and questions arise, draw on the insights of others – ask your study group, perhaps, or check in with your professor.

My word to you: You do not need to understand a lot to ask questions, but you need to ask questions to understand a lot.

NOTES

1 Mpemba, E. B.; Osborne, D. G. "Cool?" *Phys. Educ.* **1969**, *4*, 172–175. (See also an interview with Mpemba and Osborne here: https://youtu.be/dOAUdJR0SIo; Last accessed July 30, 2022.)

2 Marking, underlining, and highlighting can help with relocating points of interest in texts, even if the efficacy of such practices as study aids has been debated. Two academic papers, examples of articles that consider the question, are the following: (a) Yue, C. L.; Storm, B. C.; Kornell, N.; Bjork, E. L. Highlighting and Its Relation to Distributed Study and Students' Metacognitive Beliefs *Educ Psychol Rev* **2015**, 27, 69–78. (b) Lindner, R.; Gordon, W.; Harris, B. Highlighting Text as a Study Strategy: Beyond Attentional Focusing. Presented at: *Annual Meeting of American Educational Research Association* New York, NY, April 8–12, 1996, p. 9. A copy of the paper is available at the ERIC website with ERIC #: ED401320. https://eric.ed.gov/?id=ED401320.

3 *"The great enemy of communication, we find, is the illusion of it"* is a quote from a commentary on communication and selling failures and the need for listening by American businesses: *Fortune*, 1950, September, 77–83; 167–178 (quote on p. 174)). Similar perspectives apply, I think, to and have been articulated on knowledge and understanding. What we think we know and understand is often (much) more than we know or understand. An aside: the *Fortune* magazine article was unattributed, but a book on the subject and under the same title, by William H. Whyte, then an Associate Editor at *Fortune*, appeared in 1952.

4 Quite often, the actual answers to all or selected questions from chapters in science textbooks are provided at the back of the book. The working, however, is typically not provided there. Some textbooks have separate accompanying solutions manuals that show step-by-step procedures (the working, as we call it) for solving

problems that are included in the textbook. The suggestion here too is that you should attempt the problem seriously first before consulting the solutions manual, or other people.

5 The so-called Dunning-Kruger effect comes to mind: Kruger, J.; Dunning, D. Unskilled and Unaware of It: How Difficulties in Recognizing One's Own Incompetence Lead to Inflated Self-assessments *J. Person. Soc. Psychol.* **1999**, *77*, 1121–1134. The authors – two researchers in social psychology – found from a series of studies that (in a positive rendition that makes contact with our discussion), as we become more skilled in an area, our abilities to recognize our weaknesses also increase.

CHAPTER FIVE

Solutions in Numbers

Basic Mathematical Procedures

Many of the simplest mathematical rules that we have identified and recorded so far in human history find use in a wide
range of disciplines across the arts, humanities, natural and
social sciences, and in daily life. Algebra, unit conversions,
and discussions of ratios or rates of change (though not in the
language of calculus, and often in qualitative terms only) are
used by many people in one way or another every day. As a
science student, it is important that you are aware of and comfortable with commonly used mathematical concepts and rules,
such as those that are highlighted below. This chapter is in part
a reminder for you with hopefully some useful insights, and
it can serve too as a reference throughout your undergraduate
degree program. Many of these pointers will be reinforced later
on as well in your science training since they find extensive
application in most scientific disciplines.

It is common for students to feel less prepared mathematically than they would like to be; it is likely that many of your
classmates will feel that way initially. Mathematical knowledge
and our comfort levels with solving a given type of problem
grow, however, with time and practice. To first-year students,
taking on a stoichiometry question in chemistry, for instance,
can seem at first like a hike up a steep mountain in winter, but
the apparent demand of the question tends to shrink to something closer to a stroll in the park as the semester progresses.
Learning curves are often like that. So, take the long view –
draw on the good skills that you may have cultivated already

DOI: 10.1201/9781003263340-5

from high school and build with them a platform for the ongoing expansion in college of your analytical and mathematical skills.

Some Mathematical Reminders

(1) Algebraic Manipulations and Some Useful Math Relations

Where * represents only the + or the × operations on numbers $a, b, c, \ldots$, and the parentheses indicate what part(s) of the expression to calculate first, the following relationships hold:

$$(a*b)*c* \ldots = a*(b*c) \ldots; \text{ and } a*b* \ldots = b*a* \ldots$$

That is, the order of the numbers does not matter for a series of additions or multiplications:

$$\Rightarrow 2 + (3 + 4) = 2 + 7 = 9; \text{ and } (2 + 3) + 4 = 5 + 4 = 9.$$
$$\Rightarrow 2 \times (3 \times 4) = 2 \times 12 = 24 \text{ and } (2 \times 3) \times 4 = 6 \times 4 = 24$$

What of subtraction and division? We know that in general,

$$a - b \neq b - a \text{ and } a/b \neq b/a, \text{ unless } a = \pm b,$$

but you can always write subtractions and divisions as additions and multiplications, since:

$$a/b = a \times 1/b \text{ and } a - b = a + (-b)$$

where the usual rules for addition and multiplication apply in the same way,

$$a \times 1/b = 1/b \times a, \text{ and } a + (-b) = (-b) + a.$$

So, subtraction and division can be viewed as particular cases of addition and multiplication.

For $a \times (b + c)$, the 'a' term can be distributed into the part between the parentheses. That is:

$$a(b + c) = ab + ac.$$

For $a^{-1} \cdot (b + c)^q$, carry out the operation in parentheses first, raise that value to the power 'q,' and only then should you multiply by a^{-1} (or, equivalently, divide by 'a').

(2) TRIGONOMETRIC IDEAS

Trigonometric relationships find use across the sciences. For right-angled (or right) triangles, with angles 90°, θ_a, and θ_b, the hypotenuse (which is length c on the left in Figure 5.1) is the line across from the right angle, and we can show that $\theta_a + \theta_b = 90°$ for all right triangles, and, as Pythagoras's theorem tells us:

$$c^2 = a^2 + b^2,$$

where a and b are the lengths of the other two sides of the triangle.[1]

If we consider the angle θ_a in Figure 5.1, we can define the following functions:

$$\cos \theta_a = b / c, \sin \theta_a = a / c, \text{ and } \tan \theta_a = a / b = \sin \theta_a / \cos \theta_a.$$

And similarly, for θ_b, we can show that:

$$\cos \theta_b = a / c, \sin \theta_b = b / c, \text{ and } \tan \theta_b = b / a = \sin \theta_b / \cos \theta_b.$$

And if we directly compare the results above for θ_a and θ_b, we see that for right triangles,

$$\cos \theta_a = \sin \theta_b, \text{ and } \sin \theta_a = \cos \theta_b, \text{ and } \tan \theta_a = 1 / \tan \theta_b.$$

More generally, for any triangle, with sides a, b, and c opposite angles θ_a, θ_b, and θ_c, respectively (see Figure 5.1), we can show that

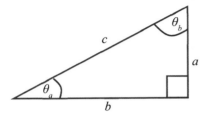
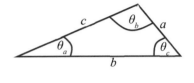

FIGURE 5.1 A right-angled and an arbitrary triangle (left, and right, respectively), with geometrical parameters labelled.

$\theta_a + \theta_b + \theta_c = 180°$ (where, for a right triangle, since $\theta_c = 90°$, $\theta_a + \theta_b = 90°$).

Additionally, we know too that for *all* triangles:

Sine rule: $\dfrac{a}{sin\theta_a} = \dfrac{b}{sin\theta_b} = \dfrac{c}{sin\theta_c}$

Cosine rule, in its various forms: $a^2 = b^2 + c^2 - 2bc \cdot \cos\theta_a$
$$b^2 = a^2 + c^2 - 2ac \cdot \cos\theta_b$$
$$c^2 = a^2 + b^2 - 2ab \cdot \cos\theta_c$$

Notice that this general 'cosine rule' has a curious resemblance to Pythagoras's Theorem, $c^2 = a^2 + b^2$ which is specific to right triangles. We will clarify the connection between them shortly.

Exercise: A claim that you can try to prove at this point, and which we will take on shortly, is that, for the special case of a right triangle, the sine rule simplifies to $\sin\theta_a = a/c$, and the cosine rule to $\cos\theta_a = b/c$.

Beyond Triangles For any given angle, θ, anywhere, you might encounter the following less commonly used definitions:

$$1 / \sin\theta = \csc\theta, 1 / \cos\theta = \sec\theta, \text{ and } 1 / \tan\theta = \cot\theta.$$

Some other useful trigonometric relationships include the following:

$$\sin(-\theta) = -\sin\theta, \cos(-\theta) = \cos\theta, \text{ and } \tan(-\theta) = -\tan\theta$$

and we can show that,

$$cos^2\theta + sin^2\theta = 1.$$

Several other interesting trigonometric relations have been established – you may have met any number of them already in high school. There is no need to know every possible trigonometric relationship, however. Understanding the key ideas and knowing the most general relationships go a very long way in sharpening your mathematical problem-solving skills. Consider, for example, the following problem:

Given that $cos^2\theta + sin^2\theta = 1$, show that $sec^2\theta - tan^2\theta = 1$

Solution

Counting Blessings

Write down any helpful piece of information from the question and any relevant bit of knowledge that you bring to the problem that might be useful in finding a solution.

This is what we know: We are told that $\cos^2\theta + \sin^2\theta = 1$, and we have definitions of $\sec\theta$ and $\tan\theta$ in terms of $\cos\theta$ and $\sin\theta$. Namely,

$1/\cos\theta = \sec\theta \Rightarrow 1/\cos^2\theta = \sec^2\theta$, and $\tan\theta = \sin\theta/\cos\theta$
$\Rightarrow \tan^2\theta = \sin^2\theta / \cos^2\theta$

So, from what we have noted so far, we can proceed now to rewrite our expression in $\cos\theta$ and $\sin\theta$ in terms of $\sec\theta$ and $\tan\theta$, and follow the logic of the process as closely as we can to get, hopefully, the result specified in the question.

Full Solution (one of many possible approaches)

Starting from $\cos^2\theta + \sin^2\theta = 1$, we can get to $\sec^2\theta$ by simply dividing through by $\cos^2\theta$.

Dividing both sides by $\cos^2\theta$ gives, $1 + \sin^2\theta / \cos^2\theta = 1 / \cos^2\theta$
Which simplifies to $1 + \tan^2\theta = \sec^2\theta$.
So, $1 = \sec^2\theta - \tan^2\theta$.

Again, there are any number of other approaches that one could take to solve this problem, but the goal is to answer the question honestly and transparently, and you can be happy if you achieve that.

Caution: If a question asks you to assume or start from $\cos^2\theta + \sin^2\theta = 1$ in proving that $\sec^2\theta - \tan^2\theta = 1$, honor that request, unless you know that it would be fine to do it the other way around, that is, to prove that $\cos^2\theta + \sin^2\theta = 1$ using the fact that $\sec^2\theta - \tan^2\theta = 1$. In general, try to do what the question asks you to do. Some professors will admire your ingenuity and independence in starting from $\sec^2\theta - \tan^2\theta = 1$ and working backward (Try it! It's basically the reverse of what we just did). Others might point out that – in their view anyway – you did not answer the question.

There are surely alternative approaches that one could take to answer this question. The emphasis here, however, is on thinking logically about the problem. Whatever strategy you might use to

take on a question like this, be sure to use the resources provided –
the information or clues in the question itself – and all that you
know otherwise to find the solution in a way that is coherent for you.

A problematic view, often espoused as a rhetorical question
about aspects of the natural sciences and mathematics, including
trigonometric rules, is: 'Why would I ever need to know that?'
And that is an unfortunate perspective, because great applications arise often from a solid grasp of fundamental aspects of
nature. For trigonometry, for instance, applications abound
in practical areas such as construction and engineering, as we
illustrate shortly with a simple model example. There is also
something to be said for appreciating the inherent mathematical
elegance of this physical universe in which we find ourselves.

Now it is true that being able to prove that $1 = \sec^2\theta - \tan^2\theta$
does not appear to be a lifesaving skill, but the ability to take
on an intellectual problem, and to work systematically, logically, and confidently through it can be lifesaving. After all,
scientists, lawyers, doctors, farmers, and even well-intentioned
politicians, all need these transferable skills for working successfully through problems. Finding cures for diseases, proving
the innocence of a convicted prisoner, setting up new strategies
for irrigation, and reducing road accidents all call for precisely
this kind of systematic critical thinking.

More Solutions: Let's show, as promised, that the sine and
cosine rules simplify to the relationships we identified previously for the special case of the right triangles.

The Sine Rule

Counting Blessings

For a triangle such as the one on the right in Figure 5.1, we
will consider a case where it so happens that θ_c is a right
angle, as shown on the left in Figure 5.1.

For any triangle: Sine Rule: $\dfrac{a}{sin\theta_a} = \dfrac{b}{sin\theta_b} = \dfrac{c}{sin\theta_c}$

For the right triangle: $\sin\theta_a = a\,/\,c$; $\cos\theta_a = b\,/\,c$ and
$\theta_c = 90°$; $\cos 90° = 0$, $\sin 90° = 1$

Solution

For the right triangle, $\sin\theta_c = \sin 90 = 1$

So $\dfrac{a}{sin\theta_a} = \dfrac{c}{sin\theta_c}$ given us $\dfrac{a}{sin\theta_a} = \dfrac{c}{1}$ and transposing

yields, as expected: $sin\theta_a = \dfrac{a}{c}$.

And we can show similarly that $\dfrac{b}{sin\theta_b} = \dfrac{c}{sin\theta_c}$ gives us, for a right triangle, $sin\theta_b = \dfrac{b}{c}$

The Cosine Rule

Counting Blessings

For any triangle: Cosine rule: $a^2 = b^2 + c^2 - 2bc \cdot \cos\theta_a$
(with two other versions shown previously)
For the right triangle: $\sin\theta_a = a / c$; $\cos\theta_a = b / c$ and
$\theta_c = 90°$; $\cos 90° = 0$, $\sin 90° = 1$

Solution

We want to show that the general cosine rule is consistent with the result $\cos\theta_a = b / c$ for right triangles. We could start with any version of the rule, and $a^2 = b^2 + c^2 - 2bc \cdot \cos\theta_a$, which already has $\cos\theta_a$ in it, seems the natural choice, but a quick diversion allows us to accomplish two objectives in this single solution.

Consider for a moment this $c^2 = a^2 + b^2 - 2ab \cdot \cos\theta_c$ version of the cosine rule. We know that $\cos 90° = 0$. So, the cosine rule simplifies readily, for any right triangle, to:

$$c^2 = a^2 + b^2$$

which is **Pythagoras' theorem!**

That result (showing the link between the cosine rule and Pythagoras' theorem) was one objective, but the main objective in this section is to confirm that the cosine rule yields $\cos\theta_a = b / c$ for all right triangles. So, let's keep going. Starting with the following version of the cosine rule,
$$a^2 = b^2 + c^2 - 2bc \cdot \cos\theta_a$$

and transposing, allows us to write:

$-2bc \cdot \cos\theta_a = a^2 - b^2 - c^2$ and therefore $\cos\theta_a = (a^2 - b^2 - c^2) / -2bc$,

But we just showed that $c^2 = a^2 + b^2$, and we will substitute the right hand side of that result into the numerator since it will help us to get rid of 'c^2' which is not in the numerator of our target (recall that we want to show that $\cos\theta_a = b / c$); and that gives us
$$\cos\theta_a = (a^2 - b^2 - \{a^2 + b^2\}) / -2bc$$

which might looks a bit ghastly, but they say the darkest part of the night is just before dawn – and here it comes, because that substitution gives us, $\cos \theta_a = (a^2 - b^2 - a^2 - b^2) / 2bc$, which we can rewrite as $\cos \theta_a = (a^2 - a^2 - 2b^2) / 2bc$ and thus, $\cos \theta_a = (-2b^2) / -2bc$. And you may have noticed already that $\cos \theta_a = (-2b^2) / -2bc$ simplifies even further to

$$\cos \theta_a = b / c,$$

which was our goal! So, there we have it!

And we can show similarly, for the same triangle, that $\cos \theta_b = a / c$.

****

Exercise: You can get to the same result by a somewhat different procedure if you start with one of the other two versions of the cosine rule, but I leave that exercise for you to take on, if you wish.

Let's return briefly, though, to the application question: For the two model houses shown in Figure 5.2, trigonometry allows us to calculate how high the vertex of the triangular-shaped roof would be, depending on how long you want 'c' to be – which ultimately means how much we are willing to spend on roofing materials. The higher the vertex gets, the large the amount of roofing material we will need.

Figure 5.2 is not a sophisticated architectural diagram, but, for that simplistic picture, you can determine the distance c, from vertex to the edge of the roof, if you know the height that you want for your roof, h, and the width of your house, x. Using Pythagoras' theorem, you can determine c for either of the two roof options (and you should convince yourself that $c = \{x^2/4 + h^2\}^{1/2}$). Once you have determined the corresponding value for c, and if you know as well the length of the building l going into the plane of the page from the front to the back of the house (since we are not building a two-dimensional house), you will be able to calculate the total surface area of the roof (i.e., $A = 2(c \times l)$). That value, A, and the area of a unit of the roofing material, can then be used to determine the minimum amount of roofing material needed for your model home. The higher h is, the longer c will have to be, and the larger the amount of materials that you will need. So, if the costs are too high for a very steep roof, one option would be to decrease c by lowering h. An alternative would be to use cheaper materials, but nobody wants a high roof that collapses a week after the house is built!

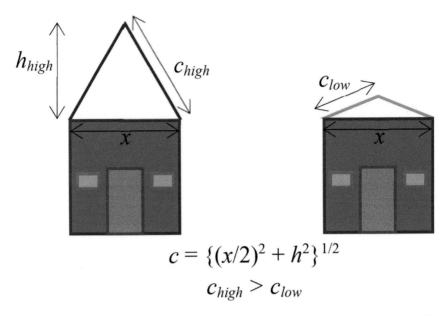

$$c = \{(x/2)^2 + h^2\}^{1/2}$$

$$c_{high} > c_{low}$$

FIGURE 5.2 Model illustrating the usefulness of trigonometric thinking, and how it finds relevance in practical situations.

(3) OTHER INTERESTING RELATIONSHIPS AND DEFINITIONS

Beyond trigonometrical relationships, you will become familiar as an undergraduate with several other mathematical relationships that find regularly use in the natural sciences. These are some of them.

$$x^a \cdot x^b = x^{(a+b)}$$

$$\left(x^a\right)^b = x^{ab}$$

$$\frac{1}{x_1} - \frac{1}{x_2} = \frac{x_2 - x_1}{x_1 x_2} = -\left[\frac{x_1 - x_2}{x_1 x_2}\right]$$

$$(a+b) \cdot (a-b) = a^2 - b^2$$

$$(a+b) \cdot (a+b) = (a+b)^2 = a^2 + 2ab + b^2$$

$$(a-b) \cdot (a-b) = (a-b)^2 = a^2 - 2ab + b^2$$

Helpful Definitions and Quantities

$$n! = n \cdot (n-1) \cdot (n-2). \ldots \cdot (1).$$

$$0! = 1$$

For a series of values $a_1, a_2, \ldots, a_n$

$$\sum_{i=1}^{n} a_i = a_1 + a_2 \ldots + a_n$$

$$\prod_{i=1}^{n} a_i = a_1 \cdot a_2 \ldots \cdot a_n$$

And it can be shown that,

$$\sum_{n=0}^{\infty} \frac{1}{n!} = \frac{1}{0!} + \frac{1}{1!} + \frac{1}{2!} + \frac{1}{3!} + \ldots = 2.71828\ldots$$

This value is given the symbol 'e.' It is an irrational number, a number that cannot be expressed as the ratio of two integers. The number 2.5, for example, is equal to 10/4. Such a ratio does not exist for e.

Exponential functions, e: The value $e = 2.71828 \ldots$ has the unique property that on a graph of y vs. x for a function of the form $y = Ae^x$, the slope at any point (x, y) on that curve is exactly equal in magnitude to the value of y at that point. Consider Figure 5.3, for instance, which is a graph of the function $y = 20e^x$. When $x = 3$, we find that $y = 401.7$ and the slope at that point on the curve is equal to 401.7 as well.

So, if we draw a tangent line at the point (3, 401.7) on the curve, as we have done in Figure 5.3, that line will have a slope that is equal numerically to the y-value at that point, which is 401.7 (see Figure 5.3). That situation, where the gradient or slope of a tangent line at a point is equal to the y-value at that point is one of the unique features of exponential functions of the form ($y = Ae^x$). That remarkable relationship between the slope and the y-value holds for every point on every curve of the form $y = Ae^x$, regardless of the value of A.

Special properties like that for a given type of function in mathematics can make them particularly useful in certain contexts. You will meet, no doubt, several other intriguing

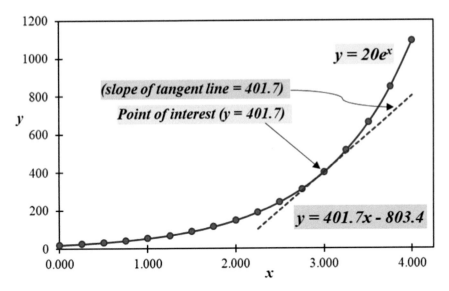

FIGURE 5.3   Graph of y = 20e^x and its tangent line (y = 401.7x – 803.4) when x = 3. At that point on the exponential curve, y = 20e^x = 401.7 and the gradient is also 401.7.

functions and will learn more about exponential functions too, as you progress in any field, from molecular physics to macroeconomics, that relies on mathematical thinking.

An equation of the form $y = e^x$ links x and y through e in the same way that the expression '100 = 10^2,' links 100 to 2 through 10. One alternative way to express the latter relationship is to say that the base-10 logarithm of 100 is 2, that is, $\log_{10}(100) = 2$. For $y = e^x$ we can also write $\log_e(y) = x$. Because of the mathematical properties of the function $y = e^x$, and because it is useful for describing phenomena across a number of disciplines, we use it and the log form, $\log_e(y) = x$, very often across the sciences. Indeed, we use $\log_e(y) = x$ so often that we almost always abbreviate $\log_e(y)$ as $\ln(y)$ – which we call the natural log – and that allows us to refer to $\log_{10}(y)$ as simply $\log(y)$ without ambiguity. Logarithms appear in expressions such as the definition of the pH of a solution,[2] and plotting graphs using the log of values instead of the values themselves can help to simplify analyses, especially where the values involved span several orders of magnitude. For example, if we performed an experiment and got (x, y) data points such as the following: (10, 1.0×10^5), (100, 1.0×10^9), and (300, 8.1×10^{10}), instead of plotting those data directly, we could plot log y vs. log x; that is, (1, 5), (2, 9), and (2.477, 10.91), which (depending crucially on the relationship between the properties represented by x and y) might allow us to produce a graph for a form of the original data that is still quite useful but employs a narrower and more manageable range of values.[3]

More Emphasis on Logarithms and Powers

The log function: If $y = 10^x$ then $\log(y) = x$.
Examples: $\log 1 = 0$, $\log 10000 = 4$, and $\log 3.163 = 0.5001$

Similarly, we can write $e^2 \approx 7.389$, so the number 2 is linked to 7.389 . . . through e. We say $\ln 7.389 \approx 2$.

The ln function: If $y = e^x$ then $\ln(y) = x$
Examples: $\ln 1 = 0$, $\ln 10000 = 9.210$, and $\ln 3.163 = 1.152$

$\log(a) + \log(b) = \log(a \cdot b)$. This is true for all logarithms, so:

$$\ln(a) + \ln(b) = \ln(a \cdot b)$$

$$\log (a)^b = b \cdot \log (a) \text{ and } \ln(a)^b = b \cdot \ln (a)$$

$$10^{\log(x)} = x, \text{ and similarly, } e^{\ln(x)} = x$$

$$\log 10^x = x, \text{ and similarly, } \ln e^x = x$$

$$\text{If } y = a \cdot 10^{-qx}, \text{ then } \log(y) = \log (a \cdot 10^{-qx})$$

$$= \log(a) + \log 10^{-qx} = \log(a) - qx \cdot \log 10$$

$$\text{So, } \log(y) = \log(a) - qx \text{ since, } \log 10 = 1.$$

And similarly:

$$\text{If } y = a \cdot e^{-qx}, \text{ then } \ln (y) = \ln (a \cdot e^{-qx})$$

$$= \ln (a) + \ln e^{-qx} = \ln(a) - qx \cdot \ln e$$

$$\text{So, } \ln(y) = \ln (a) - qx \text{ since, } \ln e = 1.$$

Linear (Straight-Line) Equations

The equation of a straight line has the general form $y = mx + b$, where m is the slope or gradient of the line and b is the point where the line intersects with the y-axis (the so-called y-intercept). That intercept is the point where $x = 0$; as you see from the equation $y = mx + b$, when $x = 0$, the 'mx' term vanishes and $y = b$.

Any equation of the form $y = mx + b$, such as $y = 5x + 2$, or $\ln(A) = \ln(A_o) - qx$ (where $A = A_o e^{-qx}$, and A_o is a constant), will give you a straight line. For the first equation, the slope of a plot of y vs. x will be 5, and the y-intercept will be 2. In the second equation, the slope is $-q$ and the intercept is $\ln(A_o)$. Notice that since the slope is $-q$, if q itself is positive, the slope will be negative, and vice versa. Four sample y vs. x plots of straight-line equations are shown in Figure 5.4.

Sometimes an equation that could take the form $y = mx + b$ is presented in a way that might seem to hide that linear character, for example, $(y - 2)/2.5 = 2x$. The latter equation is just $y = 5x + 2$, but you might have to stare at it for a bit to recognize that y vs. x would give a straight line. You might wonder, "Why would anyone want to write the equation like that?" and that's not an unreasonable question, but sometimes an equation comes out of some other context where the 'weird' form, like the ratio $(y - 2) / 2.5$, makes sense. Recognizing when you have an equation for a straight line can be enormously helpful for simplifying your analysis of scientific data and for making predictions for values of x that are of interest to you.

In other cases, it may be convenient to manipulate an equation to get it in the form of a straight-line equation, because that form can be so extremely helpful for your analysis. For example, an equation of the form $A(t) = A_o e^{-kt}$ can be used to predict the amount of a decaying radioactive element that remains, $A(t)$, after a given time, t, if the total amount that you started out with (at time $t = 0$) is A_o.[4,5] In this exponential equation, the independent variable is t, rather than x from our discussion above, and the constant in the exponent is given the symbol k rather than b, but the same rules apply. The mathematical

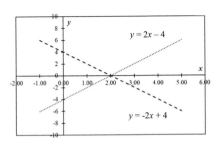 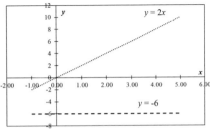

FIGURE 5.4 Sample plots of four straight-line equations. Notice that, on the left, each function is the negative of the other. On the right, cases where $m = 0$ and $b = 0$ are shown.

properties of a function are independent of actual symbols used in it or the variable (such as time or distance, for example) that is under consideration. As we noted for m and b in '$y = mx + b$,' however, the information embodied in a symbol in an equation is determined by where it appears in the equation and the phenomenon under consideration. In this particular case, k is a constant here that is directly related to the rate of the radioactive decay.

If we had some values for A for a range of t values, we could plot A vs. t and analyze the resulting graph (which would be an exponential type decay curve since, as defined, $A = A_o e^{-kt}$). But we could take another approach, which you might prefer. We could rewrite this expression in a linear form and determine A_o and k in a straightforward way. How would we do that? To obtain the linear form, we would take the natural log of both sides of $A = A_o e^{-kt}$:

$$\ln(A) = \ln\left(A_o \cdot e^{-kt}\right) \quad \Rightarrow \ln(A) = \ln\left(A_o\right) + \ln\left(e^{-kt}\right)$$

$$\Rightarrow \ln(A) = \ln\left(A_o\right) - kt \cdot \ln e$$

$$\text{So, } \ln(A) = \ln(A_o) - kt \quad \text{since } \ln e = 1$$

And once we plot $\ln(A)$ vs. t and get our straight line, all that will remain is to read the value of $\ln(A_o)$ from the y-intercept (and solve for A_o, since $e^{\ln(A_o)} = A_o$) and obtain the value of $-k$ (and get k, thus) from the slope.

Why would we wish to make the effort to determine the values of A_o and k? As we mentioned, exponential equations are ubiquitous in the sciences. We just mentioned radioactive decay, and, as another example, an equation of the form $A = A_o e^{-kt}$ arises as well in chemical kinetics. In that case, A_o would be the concentration of a certain reactant at the very start of a reaction (a fixed value) and A is the concentration of the reactant at any given time, t, as the reaction progresses. As we just saw, it's relatively cheap to determine the values of A_o and k using the linear form of the exponential equation, but the rewards are tremendous, since the values of both A_o and k are very useful pieces of information for chemists seeking to understand reactions.[6]

Quadratic Equations

Another common class of mathematical functions that you are likely to meet again is the quadratic equation: $y = ax^2 + bx + c$.

Just as straight-line equations ($y = mx + b$) have their own inherent features, such as having a positive slope if b is positive and a negative slope when b is negative, the quadratic equation has its own set of features linked to the signs and magnitudes of its three constants.

A plot of y vs. x yields a parabola, some examples of which are shown in Figure 5.5. The parabola is concave up if a is positive, and concave down if a is negative. That is a built-in feature of quadratic equations.

For any given value of x, we can show that, at that specific point on the quadratic curve, the slope $= 2ax + b$.[7] At the extremum (i.e., the maximum or minimum point on the parabola – the so-called turning point on the curve), the slope $= 0$. That is, $0 = 2ax + b$ at the extremum, such that the value of x at the turning point is: $x_{extremum} = -b/2a$.

Here, $x_{extremum}$ denotes the value of x at the extremum. For the curves shown in Figure 5.5, the extrema are all at $x = 2$, and that is as it should be. Notice that, for the upward curves, $a = 1$, and $b = -4$, so at the minima, $x_{min} = -b/2a = -(-4)/(2 \cdot (1)) = 2$, and similarly for the downward curves, $a = -1$, and $b = 4$, so $x_{max} = -b/2a = -(4)/(2 \cdot (-1)) = 2$. The value of c is irrelevant

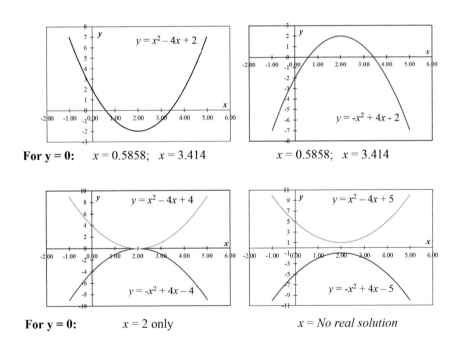

For $y = 0$: $x = 0.5858$; $x = 3.414$ $x = 0.5858$; $x = 3.414$

For $y = 0$: $x = 2$ only $x = $ No real solution

FIGURE 5.5 A few quadratic functions ($y = ax^2 + bx + c$) with positive and negative 'a' values. The curves have the same $x_{extremum}$ value but different numbers of x-intercepts (i.e., solutions for $0 = ax^2 + bx + c$) – two solutions (top graphs), one solution (bottom left), and none – that is, no real roots (bottom right).

to the value of $x_{extremum}$ for a quadratic function. Changing c simply shifts the whole curve vertically up or down, as we will shown momentarily.

We mentioned in an earlier section that the solutions of the quadratic equation $0 = ax^2 + bx + c$ are given by the formula:

$$x = \left[-b \pm \left(b^2 - 4ac \right)^{1/2} \right] / 2a \quad \textit{Quadratic Formula}$$

That equation gives the values of x at which the curve crosses the horizontal line $y = 0$. Because of the shape of the quadratic curve, there can be either two such points (see the top graphs in Figure 5.5) or only one (bottom, left) or no real solution at all (bottom, right). The actual values for x at these points where the curves cross the x-axis are given below each of the graphs in Figure 5.5. You are welcome to use the equation that we just wrote down (the quadratic formula) to check if the solutions that one gets from looking at the graphs in Figure 5.5 are actually returned by that formula.

The plots shown in Figure 5.6 are fun illustrations of the influence of changing a, b, or c only in a quadratic function. In each case, we use the function $y = 2x^2 + 4x - 4c$ (bold unbroken curves in Figure 5.6) as our reference.

As we mentioned, changing c simply shifts the whole function up or down, depending on whether c is increased or decreased. The impacts of changes to a or b are more nuanced, as illustrated in Figure 5.6, because, unlike c, which is a constant, the 'ax^2'and 'bx' terms vary with x. Two intriguing features to note are that changing only the sign of a gives an identical curve rotated (in the plane of the page) by 180° about the y-intercept, and changing only the sign of b gives a mirror image in the y-axis of the original curve (see Figure 5.6).

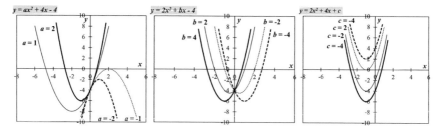

FIGURE 5.6 Graphs illustrating the influence of changes in the three constants (a [left], b [center], and c [right]) on quadratic equations. The function $y = 2x^2 + 4x - 4$ (bold unbroken curve) is common to all three panels.

Graphical Representations of Experimental Data

When we conduct experiments, we do not usually get data points that fall neatly on a straight line or smooth curve. We usually end up with graphs that look more like the examples shown in Figures 5.7. Where there is a clear trend, we may obtain a trend line – also called a line of best fit – that matches optimally the pattern evident in the data. In Figure 5.7, we show two scatter plots; one is fitted to an exponential curve of the form $A = A_o e^{-qx}$ and we show the log form as well, $\ln A = \ln A_0 - qx$.

Mathematical strategies (based on a procedure called least squares fitting) have been developed over the years to find such best-fit lines, and many common mathematical and data analysis computational programs include utilities to both plot data sets and find best-fit lines – including the equation of the line. The so-called coefficient of determination, that is, the R^2 value, is an assessment in statistics of the quality of the fit between the data and the model or type of function under consideration, with unity ($R^2 = 1$) being the ideal. That value and additional statistical data can be readily obtained from such computational programs, and it is possible in fact to test several different types of functions (straight line, exponential, polynomial, and so on) to see which one fits best, so you do not even have to presume, from observation or prior knowledge, as one might have done a few decades ago, what the best function is for the data set. Even so, although all of that work can be done for you by computers, take care to understand the

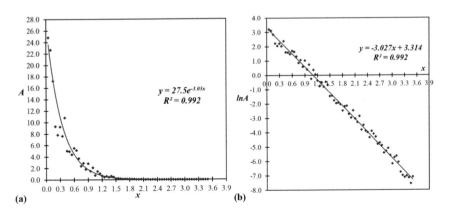

FIGURE 5.7   Sample scatter plots fitted to an exponential function (a), and a straight line (b). The two plots are related; (a) is the exponential form, $A = A_o e^{-qx}$, and (b) is the linear form, $\ln A = \ln A_o - qx$.

principles at work in the process. Taking an introductory statistics course at some point during your undergraduate experience can help with that.

Simultaneous Equations

Although modern computational tools can generate the equation for a line once the data are available, it is useful to know that we can ascertain the full equation for any straight line, for example, if we know as little as two data points that fall exactly on that line. How would we do that? It turns out that once you know the form of the equation, in this case that it is a straight-line equation with the general form $y = mx + b$, and you know that the data points (x_1, y_1) and (x_2, y_2) are both common to that straight line, we can capitalize on that shared feature of the two data points – that they both have m and b in common – to determine the values of m and b. Put another way: If $y_1 = mx_1 + b$ is true, then $y_2 = mx_2 + b$ must also simultaneously be true. So, we have two equations and two unknowns, which means we can find a solution for m and c. Indeed, the function does not need to be a straight-line equation at all. The key criterion is that we have as many data points as we have unknowns that we want to find.

Broadly speaking, simultaneous equations are functions that must all be valid at the same time.

An illustration: If I tell you that, in total, Kim and I own 12 wheelbarrows and that Kim has twice as many as I do, the claim is that both of those things are true at the same time. So, we can write two equations.

$$12 = K + I, \text{ and } 0.5 = I/K$$

(where K and I are the numbers of wheelbarrows that Kim and I have, respectively)

From the second equation, we can write, $I = 0.5 \cdot K$, and we can substitute that into the first equation to get rid of I – so that we end up with an expression in one variable, K, only. Such a substitution is valid precisely because each equation is valid as long as the other one is also valid.

And that substitution gives us:

$12 = K + 0.5 \cdot K = 1.5 \cdot K$
So, $K = 12/1.5 = 8$.   And, since $0.5 = I/K$, we can write
$I = 0.5 \cdot K = 0.5 \cdot 8 = 4$. So, **K = 8** and **I = 4**.

Similarly, if we know two data points (t_1, A_1) and (t_2, A_2) for a function of the form $A = A_o e^{-kt}$, or the log form, $\ln(A) = \ln(A_o) - kt$, we will be able to determine values for A_o and k as follows using simultaneous equations. We will show the solution for the exponential case first and then the linear form.

OPTION 1 – THE EXPONENTIAL FORM: $A = A_o e^{-kt}$

If the two data points, (t_1, A_1) and (t_2, A_2), obey the function $A = A_o e^{-kt}$ then we can write

$$A_1 = A_o e^{-kt_1} \text{ and } A_2 = A_o e^{-kt_2},$$

so $A_o = \dfrac{A_1}{e^{-kt_1}}$, and substituting into the equation for A_2 gives,

$A_2 = \dfrac{A_1}{e^{-kt_1}} e^{-kt_2}$ such that $A_2 / A_1 = e^{+kt_1} \cdot e^{-kt_2}$ and simplifying this expression gives $A_2 / A_1 = e^{(+kt_1 - kt_2)} = e^{k(t_1 - t_2)}$. And we can now solve for k by finding the natural log of both sides of the equation:

$$\ln\left(A_2 / A_1\right) = k\left(t_1 - t_2\right) \text{ from which we find } k = \frac{\ln\left(A_2 / A_1\right)}{\left(t_1 - t_2\right)}.$$

Once we have found k in this way, we can use our earlier equation for either A_1 or A_2 (see above) to find the value of A_o.

This is, $A_o = \dfrac{A_1}{e^{-kt_1}}$.

And $A_o = \dfrac{A_2}{e^{-kt_2}}$ should give the same value.

OPTION 2 – THE STRAIGHT LINE FORM: $\ln(A) = \ln(A_o) - kt$

If the data points (t_1, A_1) and (t_2, A_2) both obey the same function $\ln(A) = \ln(A_o) - kt$ then we can write,

$$\ln\left(A_1\right) = \ln\left(A_o\right) - kt_1 \text{ and } \ln\left(A_2\right) = \ln\left(A_o\right) - kt_2.$$

Solving for $\ln(A_o)$ gives, $\ln\left(A_o\right) = \ln\left(A_1\right) + kt_1$.
and substituting into the equation for $\ln(A_2)$ gives,
$\ln\left(A_2\right) = \left[\ln\left(A_1\right) + kt_1\right] - kt_2$, which simplifies to
$\ln\left(A_2\right) - \ln\left(A_1\right) = k\left(t_1 - t_2\right)$, and, as we found before,
$k = \dfrac{\ln\left(A_2 / A_1\right)}{\left(t_1 - t_2\right)}$

and by substitution into our equation above for ln (A_o), that is:

$$\ln\left(A_o\right) = \ln\left(A_1\right) + kt_1$$

we can find the value of ln (A_o) first and then calculate A_o, since, as we saw above, $A_o = e^{\ln(A_o)}$.

You can decide for yourself which of the two options that we just considered for finding A_o and k is better. The goal here was to illustrate how to solve simultaneous equations for two different types of functions. The first pair of simultaneous equations were exponential functions, and the second pair were their corresponding linear forms, and we showed, using two data points $((t_1, A_1)$ and $(t_2, A_2))$, that both approaches enable us to access the constants, A_o and k.

An Extra Example Some of the simultaneous equation problems that you will meet may be even simpler. Consider the equations $3f + 4g = 18$, and $f - 5g = -13$. You could try guessing what single pair of f and g values would be a solution for these two equations, or you could find the solutions systematically using the approach that we just employed. That is, we could tackle these simultaneous equations by transposing for f in one of the equations (perhaps the one that looks simpler to you) and then substitute that expression for f into the other equation as follows.

Transposing for f in the equation $f - 5g = -13$ gives $f = 5g - 13$ and substituting for f in $3f + 4g = 18$, gives $3 \cdot [5g - 13] + 4g = 18$. Simplifying and solving for g, we find: $15g - 39 + 4g = 18$, $\Rightarrow 19g - 39 = 18 \Rightarrow 19g = 18 + 39 \Rightarrow 19g = 57$ $\Rightarrow g = 57/19 \Rightarrow \boxed{g = 3}$.

To find f, we can substitute $g = 3$ into the transposed equation:

$$f = 5g - 13 \Rightarrow \boxed{f = 2}.$$

If we wanted to confirm our answers and put our minds at ease, we could simply substitute for $f = 2$ and $g = 3$ into the other equation $(3f + 4g = 18)$ instead to see if the left-hand side gives 18. Which it clearly does: $(3 \times 2) + (4 \times 3) = 6 + 12 = 18!$

Exercise:

(i) Going back to the type of problem that triggered our discussion of simultaneous equations, find values for m

and b for two equations of the form $y = mx + b$ that link the data points (2, 7) and (8, 19). That is, what are the values of m and b if $7 = 2 \cdot m + b$ and $19 = 8 \cdot m + b$?

(ii) What are the values of p and q, if $p + q = 1$ and

$$log\left(\frac{0.7 - p}{0.6 + q}\right) = -1?^{8}$$

Notice that the two simultaneous equations do not have to be identical in form. In the exercise, for instance, one equation is a sum and the other is the log of a ratio. All that is required is that the two parameters, p and q, each carry the same meaning in both equations and that the two expressions are simultaneously true. The solutions for the two exercise or practice questions above are given at the end of this chapter.

A Word on Matrices

Consider the two *linear* simultaneous equations that we just discussed in the worked example:

$$3f + 4g = 18$$

$$f - 5g = -13$$

An alternative way by which such a set of functions can be written down and solved is by using matrices (i.e., matrix algebra). That approach allows us to write, for example, the single equation $3f + 4g = 18$ as follows.

$$(3 \quad 4)\begin{pmatrix} f \\ g \end{pmatrix} = 18$$

The rules for matrix multiplication are that, going from left to right, we multiply terms in rows with corresponding terms in the columns and add the resulting expressions. In this case, that would give, as we expected, $3 \cdot f + 4 \cdot g = 18$. One consequence of this way of writing down an equation is that the first matrix must have as many columns as the second one has rows. For a pair of simultaneous equations, this happens automatically. For the two equations that we wrote down at the start of this section, for example, the corresponding matrix form is,

$$\begin{pmatrix} 3 & 4 \\ 1 & -5 \end{pmatrix}\begin{pmatrix} f \\ g \end{pmatrix} = \begin{pmatrix} 18 \\ -13 \end{pmatrix}.$$

And we can abbreviate that expression as $M\begin{pmatrix} f \\ g \end{pmatrix} = \begin{pmatrix} 18 \\ -13 \end{pmatrix}$.

But our real objective is to find f and g. So, how do we do that using matrices? I thought you'd never ask!

To solve for f and g, matrix algebra affords us a few options.

One approach that we might follow is called Cramer's rule. To do so, we have to define what we refer to as a *square matrix* and introduce the concept of the *determinant* – a value that is characteristic of a given square matrix.

A square matrix is a matrix that has the same number of rows and columns, such as the following $n \times n$ cases, $\begin{pmatrix} a_1 & b_1 \\ a_2 & b_2 \end{pmatrix}$

and $\begin{pmatrix} a_1 & b_1 & c_1 \\ a_2 & b_2 & c_2 \\ a_3 & b_3 & c_3 \end{pmatrix}$, where $n = 2$ and 3, respectively.

The determinant (denoted by D or $|M|$) of a 2×2 matrix M, is the number obtained conveniently by multiplying the terms that are diagonal to each other, that is, '$a_1 \cdot b_2$' and '$a_2 \cdot b_1$', in the

matrix $\begin{pmatrix} a_1 & b_1 \\ a_2 & b_2 \end{pmatrix}$ and subtracting the latter from the former:

$|M| = a_1 \cdot b_2 - a_2 \cdot b_1$.

Cramer's rule tells us that where $\begin{pmatrix} a_1 & b_1 \\ a_2 & b_2 \end{pmatrix}\begin{pmatrix} f \\ g \end{pmatrix} = \begin{pmatrix} c_1 \\ c_2 \end{pmatrix}$

we can solve for f and g by computing three determinants using the values in the columns in the equation:

$$D = \begin{vmatrix} a_1 & b_1 \\ a_2 & b_2 \end{vmatrix}; \quad D_f = \begin{vmatrix} c_1 & b_1 \\ c_2 & b_2 \end{vmatrix}; \quad D_g = \begin{vmatrix} a_1 & c_1 \\ a_2 & c_2 \end{vmatrix}$$

and the values of the unknowns are

$$f = D_f / D \text{ and } g = D_g / D.$$

Applying that approach to our problem, $\begin{pmatrix} 3 & 4 \\ 1 & -5 \end{pmatrix}\begin{pmatrix} f \\ g \end{pmatrix} = \begin{pmatrix} 18 \\ -13 \end{pmatrix}$,

we can show that f and g are given by the following ratios of determinants:

$$f = \frac{\begin{vmatrix} 18 & 4 \\ -13 & -5 \end{vmatrix}}{\begin{vmatrix} 3 & 4 \\ 1 & -5 \end{vmatrix}} = \frac{-38}{-19} = 2 \quad \text{and} \quad g = \frac{\begin{vmatrix} 3 & 18 \\ 1 & -13 \end{vmatrix}}{\begin{vmatrix} 3 & 4 \\ 1 & -5 \end{vmatrix}} = \frac{-57}{-19} = 3$$

Another approach that is often taken in traditional introductions to matrix algebra is included in Appendix I. That approach leads us to the same place but directly invokes concepts such as the identity, adjoint, and inverse matrices that you may learn more about in courses on matrices and linear algebra.

For solutions to two simultaneous equations, the matrix route might look excessive. The strategy that we outlined previously of substituting from one equation into the next and so on seems to be simpler and more straightforward – even for cases that involve simple exponential and other non-linear functions. Yet, matrix algebra proves to be exceptionally useful for solving larger systems of simultaneous linear equations. Finding determinants for square matrices with $n \gg 2$ gets very involved – we took some shortcuts (specific to $n = 2$) around the general rules that apply to any n. The rules are well-established in mathematics and can be applied systematically for any n, but finding solutions by hand gets horrendous fast as n increases. Thankfully, though, computers allow us nowadays to find or confirm almost any solution that we could want, even for rather large values of n.

For this discussion, our key goal is, however, to introduce the idea of matrices and emphasize the usefulness of matrix algebra. This brief excursion will free you, I hope, as the need or wish arises, to learn more and to consider matrix approaches if a problem seems to call for it. For more on how matrices and other neat but ostensibly abstract mathematical tools find cool applications in the natural sciences, see, for example, a non-technical commentary mentioned in the chapter endnotes[9] on mathematics in biology.

On the Shapes of Things
CIRCLES, CYLINDERS, AND SPHERES

For any *circle* (Figure 5.8), the ratio of the perimeter (i.e., circumference, C) to the diameter, d, of that circle is a constant. That is,

$C \propto d$ and the constant of proportionality is given the symbol π. So, $C = \pi \cdot d$, and we find that for all circles, $\pi = 3.14159 \ldots$

But we know that – by definition – the diameter $= 2r$ (where r is the radius of the circle), so we can write,

$$C = 2\pi r.$$

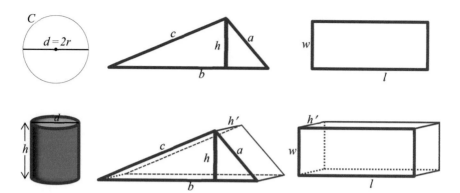

FIGURE 5.8   Key parameters for some basic shapes you will encounter.

And for any given circle of radius r, the area, A, is:

$$A = \pi r^2.$$

A *cylinder* of height h may be seen as a rectangle curved so that two opposite ends, each of equal length h, just touch. The top and bottom of the cylinder are circles, each of circumference $2\pi r$ and area πr^2. For such an object, therefore;

Surface Area $= h\cdot(2\pi r) + 2(\pi r^2) = 2\pi r\cdot(h + r)$ and Volume $= \pi r^2 h$

A *sphere* of radius r can be seen as a surface generated by a 180° rotation of a circle of radius r about an axis that passes through the diameter of the circle. For such an object :
   Surface Area $= 4\pi r^2$ and Volume $= (4/3)\cdot\pi r^3$

TRIANGLES AND (TRIANGULAR) PRISMS

Consider a triangle with sides of length a, b, and c, and a maximum vertical height, h perpendicular to the base of the triangle (Figure 5.8):
   Perimeter $= a + b + c$, area $= (h \times b) / 2$, and the volume of a triangular prism $= (h \times b \times h') / 2$, where h' is the length of the rectangular face of the prism projecting into or out of the plane of the page (Figure 5.8).

RECTANGLES AND CUBOIDS

For any *rectangle*, the perimeter $= 2l + 2w$ (where l is the length; w is the width), area $= l \times w$, and the volume of a rectangular cuboid $= l \times w \times h'$ (see Figure 5.8).

Layer upon Layer

As we said at the beginning of this chapter, mathematical knowledge and one's comfort level with solving a given type of problem builds with time and practice. Do not let the typical anxieties about mathematical demands that so many students feel early on in college science cause you to miss the rewards that will follow from persistence and your maturing practice of good academic skills. In time, as your knowledge base expands, and your experience and confidence grow, you will be much more at home with applying mathematical solutions to problems in your courses or research activities. Another way to say this is that you, like your professors, should have a *growth* mindset.[10] Focus on your intended personal, academic, and professional destination and how you can get there, rather than defining yourself (or allowing yourself to be eternally defined) by deficits – what you have not managed so far to attain. Prevent biased negative perceptions – be they yours or those of others – about your ability and ambition from limiting your aspirations or stifling motivation. Do not build or nurture mental barriers between you and your potential achievements. You and those who teach and support you should expect continued intellectual growth and development, which will be achieved by the work you will be doing, and through the instruction, advice, and mentoring that you will receive.

Each new piece of mathematical insight or skill will build, layer upon layer, a strong foundation for potential application in one course or another in college and potentially in remarkable ways at other stages of your life as well.[11] The mathematical principles and approaches highlighted here are by no means exhaustive. Calculus, probability and statistics, and vectors (see Appendix II; which tends to evolve into a discussion of matrices) are some of the other areas in mathematics in which science majors will develop some expertise during their undergraduate training. Calculus finds important use across numerous areas of study well beyond the natural sciences,[12,13,14] even if its standing as the dominant content in introductory college mathematics has been challenged.[15] You may not see it played out in the classroom, but academics (with equal fervor in every discipline, I think) have heated discussions about what should be added to, magnified or diminished in, or fully booted from the curriculum. Probability and statistics are also ubiquitous, finding prominent applications in areas such as biostatistics, econometrics, quantum mechanics, and statistical mechanics. The connections of probability and statistics to data science and machine learning will only continue to elevate our appreciation of the tools that they provide and the associated ways of thinking.

Faculty members who teach mathematically intensive advanced courses in the sciences will continue to debate the list and depths of topics covered, and thus the level and type of mathematical training that should be required for students entering their courses. Those debates are motivated by several considerations, including the limits of instructional time in a semester or academic year, the availability of increasingly powerful computers and specialized algorithms that allow us to perform accurately and rapidly (even on the fly in the classroom) ever more copious amounts of complicated mathematical operations. Other considerations include the desire to incorporate materials from more recent advances in the discipline, the growing need for programming skills in many areas, and evolving demands on the modern science major entering graduate school or the increasingly sophisticated job market.

What is not in question, however, is the need for students to develop, through all of the stages of the undergraduate curriculum – beyond the requisite understanding of the core content and language of the discipline – the ability to think critically, solve new problems, and to think in new ways about old problems. And these skills are reinforced in mathematical training by thinking intentionally, not only about the answer to a problem on an assignment, but by reflecting too on the logic that yielded that solution and how similar ways of thinking might open up opportunities for solving other problems.

A Fun Illustration from Shapes

What area can we encompass with a 20 m long string? We know from mathematics that a *square* with a perimeter of 20 meters (with each side 5 m long), has an area of 25 m², and we can show as well that any other *rectangle* that you can make with the *same* perimeter will have a *smaller* area. For example, a rectangle with two sides 6 m in length and the other two sides 4 m in length will have the same perimeter (20 m) as our square, but a different (smaller) area: 6 m × 4 m = 24 m². And what of a *circle* with the same perimeter of 20 m? We can show that since the perimeter (circumference) of the circle is $2\pi r = 20$ m, the radius is $r = 20$ m $/ 2\pi = 3.183$ m. And since the area for a circle is πr^2, we can write: $Area = \pi \cdot (3.183)^2 = 31.8$ m. So, the circle has an even larger area than the square or any other rectangle of the same parameter. In fact, a circle has the largest area of any two-dimensional shape of a given perimeter. So, what? How do the facts on area vs. perimeter connect to the current discussion? Those observations may be nothing more to some than curious or interesting claims to remember for a test. Yet, another student, a budding architect perhaps, might start to envision how those facts could play into the shapes and designs of buildings, and an aspiring entrepreneur in the same math class might start to rethink a creative packaging idea – switching from a rectangular ice-cream cup to a circular one to ensure customers get more for the same perimeter (if not the same price)! The value of mathematical thinking, regardless of your disciplinary interest or professional aspirations, will be limited only by imagination.

STAY THE COURSE

You will gain some of your mathematical training (i) in a systematic way through math courses that you might volunteer or be required to take as prerequisites in college, and (ii) in a more *ad hoc* fashion based on skills you will need to learn and apply in studies beyond formal mathematics courses (in areas such as chemistry, computer science, economics, physics, and any number of other disciplines). In any case (and in good time – even if others seem to be ahead to start with, especially in first year courses as people come in from many different high school backgrounds), the mathematical competencies that you will need will come. They will be facilitated by faithful, if also incremental, steps forward, and with the support of your peers and instructors; and the more you learn, the more you will be able to read, comprehend, and apply mathematical ideas and logic independently, happily, and successfully. So, stay the course.

NOTES

1 In some disciplines, such as mathematics, and for various topics across the natural sciences, angles are often expressed in radian units. Let's say you have a circle of radius r. If you traverse a distance equal to exactly r on the perimeter of that circle, the corresponding sector angle (defined by the center of the circle, and the points on the perimeter where you started and where you stopped) is 1 radian. If you traversed instead exactly halfway around the circle, the corresponding angle (which is 180°) is ~3.1416 radians – more exactly, π radians. So, 180° = π radians, 90° is $\pi/2$ radians, and so on. An advantage of radian units is that it directly links angle to the corresponding distance traveled (in radians or 'radius' units) on the circumference of the circle of that

radius. We stick to degree units here, however, since degrees are more commonly used in many other areas in the sciences and in other areas of daily life. Many students entering college will have a more developed intuition for degree units.

2 The pH of a solution is the negative of the log of the hydrogen ion concentration, typically denoted $[H^+]$, or more realistically $[H_3O^+]$, in an aqueous solution. That is, pH = $-\log[H^+]$. Acids have a higher $[H^+]$ and bases have lower $[H^+]$ than pure water at the same temperature. Yet, H^+ concentrations for solutions can be really very small values in moles/dm^3 (M) units, such as 1.00×10^{-12} M, and one common way to get around using those inconvenient numbers is to use the logs of the values instead. So rather than saying $[H^+] = 1.00 \times 10^{-12}$ M, we say the pH is 12, since $-\log(1.00 \times 10^{-12}) = 12$. Note that pH values do not have units.

3 $\log(10) = 1$, and $\log(1.0 \times 10^5) = 5$; $\log(100) = 2$, and $\log(1.0 \times 10^9) = 9$; $\log(300) = 2.477$, and $\log(8.1 \times 10^{10}) = 10.91$.

4 $A(t)$ here does not mean A multiplied by (t). It is a way of affirming that A explicitly depends on time, t. In the same way, we sometimes express functions in the form $f(x) = mx + b$ or $f(t) = A_o e^{bt}$, for example.

5 The quantity A_o is not a variable, it is a constant in the equation, and the subscript 'o' is there to remind us that it is the fixed amount with which we started at time $t = 0$.

6 For such a reaction, k is a value called the rate constant. It is a quantity that relates the rate of the reaction to the concentration(s) of the reactant(s) in a chemical reaction. You will likely learn about rate constants and reaction rates if you take an introductory chemistry course in college.

7 Expressions for the slope of any curve, for any x, can be obtained using calculus. We will not prove that here, but a foundational calculus course will enable you to become comfortable with the meaning and use of differentiation and how it is used to obtain an expression for the slope at any given point on a curve for which the function is known.

8 The solutions are (i) $m = 2$, and $b = 3$; and (ii) $p = 0.6$, and $q = 0.4$.

9 Cohen, J. E. Mathematics Is Biology's Next Microscope, Only Better; Biology Is Mathematics' Next Physics, Only Better *PLoS Biol.* **2004**, *2*, e439 (2012–2023).

10 *Mindset: The New Psychology of Success*, Carol S. Dweck, Ballantine Books; Updated Edition, 2007.

11 For an interesting case, see: *How a retired couple found lottery odds in their favor*, Jon Wertheim, CBS News, January 27, 2019: www.cbsnews.com/news/jerry-and-marge-selbee-how-a-retired-couple-won-millions-using-a-lottery-loophole-60-minutes/ Last accessed March 5, 2022.

12 Lax, P. D. *In Praise of Calculus*, p. 1–2 in MAA Notes #6 (The Mathematical Association of America). Toward A Lean and Lively Calculus: Report of the Conference/Workshop to Develop Curriculum and Teaching Methods for Calculus at the College Level, Tulane University, January 2–6, 1986, Editor: R. G. Douglas.

13 Douglas R. G. *The Importance of Calculus in Core Mathematics,* pp. 3–6 in MAA Notes #6 (The Mathematical Association of America). Toward a Lean and Lively Calculus: Report of the Conference/Workshop to Develop Curriculum and Teaching Methods for Calculus at the College Level, Tulane University, January 2–6, 1986, Editor: R. G. Douglas.

14 Bressoud, D. M. Why Do We Teach Calculus? *Amer. Math. Monthly,* **1992**, *99*, 615–617.

15 Douglas R. G. *Introduction: Steps toward a Lean and Lively Calculus*, pp. iv–vi in MAA Notes #6 (The Mathematical Association of America). Toward a Lean and Lively Calculus: Report of the Conference/Workshop to Develop Curriculum and Teaching Methods for Calculus at the College Level, Tulane University, January 2–6, 1986, Editor: R. G. Douglas.

CHAPTER SIX

Practical Solutions

Science in the Laboratory

Why Experiments Matter

A good experiment can cause you to leave the room understanding things that you did not even know were there to be understood. Experiments can change your outlook on things. See undergraduate science laboratory activities, therefore, as opportunities for new insights into and discoveries about the nature of the universe. See them too as apprenticeships, formative experiences that are preparing you – through technical and disciplinary training and practice in applying transferable critical thinking skills – to make your own scientific discoveries someday.

The basic approach that scientists use in the practice of experimental investigation is called the scientific method.[1,2] Put simply, the method offers an orderly general mode of operation for an inductive or empirical study. It's not a guaranteed path to extraordinary discoveries (see n. 1). It's simply an organized way to possibly uncover new knowledge by systematic investigation.

Very briefly, the scientific method starts with a question that you have about the physical world, requires the logical design of an experiment to answer that question and the careful acquisition and analysis of experimental data, and it culminates with the reporting of your results and any conclusion (including any success or failure that you experienced in addressing the question). The expectation is that all conclusions will follow systematically from the evidence; a logical thread linking

DOI: 10.1201/9781003263340-6

evidence to conclusions will allow others to follow, appreciate, and build on your work. Often, based on prior knowledge, including insights from relevant physical laws, scientists will posit an explanation or answer – a hypothesis – to go along with the question at the start of the investigation. Hypotheses are not always necessary.[3] If your question is, for instance, "What route do birds take to travel from Falmouth, Maine, to Inagua, Bahamas, and back to Falmouth?" one might reasonably (oversimplifying a bit here) strap position trackers and cameras to the birds, wait, and record the migration data with no specific hypothesis at all about bird travel. You are assuming of course that birds will behave in the same way whether they have the monitor on or not, and that's a separate consideration; you'd be wise to ensure that the monitor is as sturdy, yet light, small, unrestrictive, and inconspicuous as possible.

If your question is, however, "Why do birds travel from Falmouth, Maine, to Inagua, Bahamas, and back to Falmouth?" You might posit that it is because of temperature or weather changes or bird-food insecurity in Maine, and your hypothesis (or set of hypotheses, see endnote 3) will be decisive for how you proceed with setting up your experiment. In such cases, the hypothesis is crucial for the design of experiments.[4] Indeed, the goal of the experiment will be to test the hypothesis. The goal is never to prove the hypothesis, however; the *evidence* should always lead, not your hopes, dreams, or biases. What you will learn in undergraduate teaching laboratories, and hopefully too in a mentored research experience, will be how the scientific method is applied in practice to tackle interesting problems and to answer significant scientific questions in your area(s) of interest.

The instructions for any upcoming laboratory exercise will usually be provided by a laboratory manual developed specifically for the course. When you enter the laboratory, you will be happy if you prepared well. Otherwise, the experience will be unnecessarily taxing and far less pleasant than it could have been.

As you prepare, read the experimental procedures (usually written in the laboratory manual) as thoroughly as possible. Some courses will come with mandatory pre-laboratory exercises or questions (so-called 'pre-labs') that you will be required to complete before you show up for the (usually weekly) laboratory activities. Those preparatory assignments make the actual laboratory session much more meaningful and enjoyable – they should be taken seriously, and we will say more about them in a later section.

Once the laboratory activities get underway for the day, you should – a reminder here – see the experience as another chance to learn more about this intriguing universe in which we live. Focus on the joy of being able to test and to (dis)prove ideas that you only heard about before. Embrace the active learning experience. Do not be overly stressed about handling new equipment, or by the thought that you might not know everything you will need to know to complete the experiment successfully that day. If you can manage it, relax, be positive, follow the experimental procedures in an orderly and safe way, and dive in – keeping the laboratory rules in mind: not eating, drinking, or playing in the laboratory, handling chemicals and equipment safely and with appropriate personal protective equipment, and so on (see Appendix III on safe problem-solving).

Take good notes during the experiment on what you are doing and observing. You do not need to rewrite the procedure for the experiment unless you are asked specifically to do so, but you might write down, for example, that you or the team you are working with, "Completed step #1 as directed successfully." Be sure to record as well any significant exception or deviation from the originally prescribed procedure issued for the experiment. Make a note, for instance, of the fact that, "Step #2 took three times as long as the directions said it should, even though we did not deviate from the procedure." And you are free to add, "We are not sure at all what happened – should check with the lab T.A." Such a note might remind you, later on, to send a message to the teaching assistant (T.A.) about your observation, and to mention in your final report too perhaps that the directions might be understating the time required for Step #2.

You will usually be expected indeed to write a report after you leave the laboratory, a document that will include an analysis of the results you obtained during the experiment. That report may eventually be submitted for grading and the professor should get the graded outcome back to you as soon as possible. Once your laboratory report has been graded and returned, be sure to check for successes and any mistakes or errors that you made. This step is very important for laboratory experiments, because some skills, such as stoichiometric calculations in chemistry, or making accurate volume and other physical measurements, and the preparation of aqueous solutions will be called on repeatedly in future experiments. So, if you can find ways to improve on your laboratory skills early on (hence the need for early feedback), those lessons and improvements will be gifts that will keep on giving.

Start working on your laboratory report as early as possible after you leave the laboratory. The experiment will be fresh in your mind, and if your notes have gaps or errors in them from what you wrote during the laboratory activity itself, it may be easiest to fill in those gaps or identify the errors promptly after completing the experiment.

Many courses require students to work in groups, even for very short periods. Laboratory classes, for example, may require students to work in pairs, and your so-called 'lab partner' or 'lab mate' in your first laboratory course may be someone who is culturally different from you – having a different background, even a different nationality, and different ways of being and doing. Embrace such interactions as opportunities to grow in your ability to work and learn with others who are growing in that same way toward you as well. Working with a lab partner effectively can help you to complete laboratory activities more efficiently and accurately. Sometimes you and your lab partner will work separately on different parts of a scientific investigation, and that kind of division of labor can help you to get a lot of good work done. Sometimes you will control and monitor the process as a system heats up and your lab partner will focus on writing down the temperature readings every 30 seconds (even in modern laboratories some things are still manual). But sometimes you will work together – you will both review the final graphs or tables or reread your group's final big report, and having a second pair of eyes can be enormously helpful in those cases.

As with college roommates, however, if lab mates are hostile, threatening, racist, or behave in other unacceptable ways, report your concerns to the relevant authorities – the laboratory assistant and instructor, for example – and request a new lab partner. You will never find another person who aligns perfectly with your likes and dislikes, your personal philosophies, and every aspect of your worldview; learning to thrive in diverse communities, with various people with myriad perspectives on the way of things, all without being forced to lose yourself, is important. But if you find that an interaction is tending toward the abusive (in some of the ways mentioned above), do not tolerate it – ever. That said, I have rarely seen anything close to that kind of situation. Lab partners are usually able to develop very good working relationships, and even strong friendships beyond the life of the course in some cases. As with most other interactions, however, building a good rapport early on and preserving mutual respect and empathy will help to ensure a successful partnership.

A laboratory course can change your life. It can show you the power of a subject and stir you to redirect your professional aspirations to work in that particular discipline. Laboratory activities can engage touch, sight, smell, and hearing, and mind and heart, in ways that conventional lectures do not. You might meet, too, some workshop-style courses, where lecture and laboratory exercises merge into a single active-learning class-room experience. Colors change, a solid finally melts, a limb is regenerated, an odor emerges from a reaction, a projectile lands precisely where your mathematical analysis predicted . . . and you – like centuries of thinkers before you – are captivated by the thrill of it, questions firing off in your mind, the irresist-ible 'Hows' and 'Whys' of things. Again, embrace laboratory experiences!

It can be hard to keep up when the semester is in full swing. You have a test, a meeting, friends to catch up with, a paper to write, and that laboratory report, but you can manage through those challenges. Many with skills no sharper than yours have done it quite successfully.

Approaching Laboratory Activities

The key to an enjoyable experience in the science laboratory is to see it as a few hours of active exploration and inspiration. *Be prepared to be amazed* should be your theme for each labora-tory session. If you have instead a theme like *O God, help me to not make any mistake today* you are setting yourself up for a tense and unnecessarily miserable experience.

Pre-laboratory assignments are designed to provide you with an introduction to the key concepts and mathematical skills that will be needed for the next experiment. Completing them can help a lot with your comfort, efficiency, success, and happiness when you are finally in the laboratory, because they make things (the laboratory equipment and procedures, for example) look and feel more familiar. Things can get hectic as the semester progresses. Staying on schedule is not easy, time management can be a challenge in college, but the greatest benefits of pre-lab assignments will ensue when they are com-pleted in advance rather than in a rush a few minutes before the laboratory session begins.

As we mentioned in an earlier section, preparation sets up a virtuous spiral where the joy of learning is magnified by prepa-ration and that joy encourages preparation for the next class, and so on. Pre-laboratory reading and practice are only part of what will prepare you for success in the laboratory. Reading the textbook and engaging in the classroom (for courses with both lecture and laboratory-type components) will all contribute to excellence in the laboratory. Ultimately, assuming that the teaching is excellent and the environment is supportive, your preparation, organization, and motivation can make the difference between looking forward to the next experiment (or the next class in general) and dreading it.

Here are a few tips for making the most of your laboratory experience.

Insist on High Standards of Logic and Reasoning

Laboratory experiments usually come with challenging questions about or based on observations made during the experiment. You might even have to do some data analysis in that process and generate graphical representations of data using in-house computational codes or commercial software.

When you settle in to answer such questions, do not simply guess answers and write them down with no logical basis for your response. Even if your explanation for an observation turns out to be incorrect, it should not be arbitrary. For instance, let's say that in step two of an experiment you poured a colorless solution of compound B into a container with a colorless solution of compound A to form a new blue solution. And let's suppose that one of the questions on your laboratory report sheet is, "Why did mixing the two colorless solutions give a blue mixture at step two?" You may not be sure why, but it would be unacceptable to say, "Because we added B to A." It is expected that you will answer questions as closely as possible to (if not above) your current level of training, and that answer would clearly be well below par in a first-year college course. The question is seeking a level-appropriate explanation, not just a statement of fact. That "we added B to A" is a fact – if B and A were not mixed, you would not have had a mixture – but stating that fact offers no insight or logical handle on the actual question – that is, the origin of the blue color. The statement, "Because we mixed B and A" says something; it reminds us of the experimental procedure, but it explains nothing. An answer like, "The solution turned blue because compound A reacted with compound B, to form two new compounds C and D. Compound C is known to be colorless, but D is known to be blue in solution" would be a much better answer. Even if the compounds formed by the reaction were in fact E and F (two different products from C and D), and both E and F are blue, your answer shows that you really thought about the problem. Sure, your prediction was incorrect that time around, but your approach to thinking about the problem was rational.

BE WILLING TO THINK INDEPENDENTLY AND TAKE ON NEW CHALLENGES

Success in the laboratory relies on your willingness to take on challenges and to deploy your mental powers independently to tackle unfamiliar problems. Yet the teaching laboratory is also a collaborative space and, for some problems, it may be a hint from your laboratory partner or an instructor that nudges you finally toward a solution. For some problems, the answers will be embarrassingly obvious to you; others will make you sweat a bit in the hunt for solutions. If the rules allow for it – which they should in most undergraduate laboratories – do not feel that you are bothering the instructor by sharing with them that you are making no inroad into a problem and would like to share the gist of your approaches and benefit from their insights.

Your knowledge, clear thinking, patience, and pluck will get you to success most of the time, but your professor is a mentor for you in problem-solving, a guide, and a cheerleader too. So, even in cases where you are expected to work independently, it may be fine to ask your professor if you are moving in the right direction. That, of course, is usually NOT the case for formal assessments, such as traditional in-class or take-home tests or exams, but for formative activities (working on practice questions, problem sets, laboratory assignments, and so on), such consultations are encouraged. If you work on enough problems, you will find that luck comes to the rescue sometimes, and that, inevitably, you will miss the mark every now and then – all good scientists do; problems resist our efforts at solutions – but persist. Press on. Do not sacrifice to fear and intimidation the gifts that determination will usually have in store for you.

AN APPRECIATION OF ERRORS

You are likely to have detailed discussions early on in your undergraduate laboratory courses about error analysis – uncertainties and errors (random and systematic), accuracy and precision, error propagation, least squares fitting, and so on. And the treatment of those topics will become more quantitative and rigorous as you progress in college. A discussion of error analysis is not attempted here, but let us close this section with a few comments on how to think, in broad terms, about sources of error in science laboratories and their impact on the quality of your results. The technical term *'sources of error'* refers to unintentional or unavoidable negative influences on the quality of your experimental data. Sources of error include any inherent defects or limitations to equipment or instruments used to make measurements in the laboratory, as well as flawed practices and human limitations in making measurements. Examples include a scale that needs to be calibrated or is rusting (whether we are aware of those defects or not) and that gives slightly wrong masses for your samples, the limited number of significant figures available from a given instrument, parallax errors,[5] or an odd mistake that you made by tilting the measuring cylinder while reading a volume from it.

When you report your results, you will normally be required to comment on such sources of error, so be on the lookout for them as you work. You cannot account for unknown sources of error (such as unseen corrosion inside a balance), but work to minimize those sources of error over which you have some control – such as parallax errors – and report errors that are inherent to laboratory equipment and experimental procedures, such as the limited number of significant figures available from an instrument. With minimal training, you will be able to tell the uncertainties inherent in experimental measurements due to the specific number of significant figures available from an instrument, and you will even be able to assess the impact of those uncertainties on quantities derived from them (a process called error propagation or the propagation of uncertainty). Guidance for handling numbers – writing values in standard form, determining the numbers of significant figures in a value, rounding numbers, and reporting the correct number of significant figures after a mathematical operation like addition or multiplication, for example – is not outlined here, but will usually be provided in laboratory manuals or handouts and often in the first chapter (or as a dedicated appendix) in introductory college textbooks in the sciences.

The fact that you poured alcohol rather than water into a solution by mistake is not a 'source of error' in the classical sense. The mistake might have rendered the whole experiment meaningless for answering the original experimental question. If such a mistake occurs, it should be mentioned to the laboratory instructor or addressed, perhaps, by repeating the procedure. If it is too late to repeat the experiment or overcome the mistake in some other way, you should point out the error in your laboratory report, and you are free as well to posit, based on your knowledge of the topic and understanding of the specific experiment, what you would have expected to happen if you had followed the directions. As always, however, the first thing to do in response to unusual developments in the science laboratory is to check with your instructor. They have seen (and done) a lot, and are likely to be much more understanding in some cases than you think. So, do not feel that a mistake or two puts you out of the running for a high mark in the course or a passing grade for that specific laboratory exercise.

If it's your lab partner who made the critical mistake instead of you, try to be gracious there too.

ANOTHER SUGGESTION TO KEEP IN MIND

In a qualitative account of possible sources of error in your experimental results, do not simply say, for example, "Maybe some of the solution spilled onto the table as I poured it from one beaker to the next." Or, "Maybe we made a math error." If you saw evidence of a spill or know that you made an error in your calculations that affected how you carried out an experiment and the eventual results, say so directly, and then discuss logically the potential impact of such a mistake on the accuracy of your final result. If you can go back to your notes and see where an error was made in a calculation, you may be able to correct it, if you have time, or give a full and honest assessment of the consequences of that error on your observations and the meaning of your results. If you miscalculated how much of one compound to put into your reaction vessel, for instance, that might explain why you did not see the bright purple color that others in the laboratory observed. Having identified that error, you will be able, in retrospect, to provide a fulsome, transparent, and honest explanation in your report of what happened.

THE UNKNOWN POSSIBILITIES

In general, wild guesses cannot properly account for unusual phenomena or deviations from expected results. Do not cite as either a mistake or a source of error a remote possibility for which you have no evidence, such as "I think the sample *might have* been impure" or "The earth's magnetic north *might have been* constantly shifting." Maybe the sample was impure, maybe the pole is shifting, but with no evidence that such claims are true, the comment is unhelpful. You could cite such remote possibilities in your report, however, as aspects that you think are worthy of further investigation (based on a hunch you have, an alternative hypothesis, or basic curiosity).

That would be fine. Some professors might even create a space outside regular laboratory sessions for you to test your 'strange' idea.

Ethical Engagement

Going back to an issue we hinted at previously. What would you do if you made a potentially devastating mistake that could cost you a good grade or you got some results from your experiment or assignment that are wildly different from what you expect or think the professor expects? What should you do? In all cases, it is required of scientists to preserve the highest ethical standards in conducting and reporting their work. Take responsibility for your mistakes, and report results that are faithful to what you observed or calculated. Do not invent or falsify data to suit (presumed) expectations or for any other reason. Do not plagiarize. And guard against other temptations (or efforts by peers) to deceive or delude instructors or any other audience about the nature or quality of your study and its results. If your solution did not undergo a color change, do not claim or pretend that it did. Put another way, if your professor made a mistake and gave you the wrong solutions such that no color change should have occurred, do you want to be one of those students who claim to have seen a change because they knew that was the expected result? You will naturally be worried that the laboratory instructor, for example, will be disappointed or displeased and deliver a disastrous grade upon reading the report, but you might be surprised how impressed the professor will be if you assess the situation honestly and plainly. You might even say, "My lab partner and I expected a color change based on our understanding of the experiment, but we did not observe any. We cannot find any good reason for that outcome, and would be happy to discuss it with you in case we overlooked something or there is some other explanation." There is no way to guarantee what kind of grade you will receive if you really did make a momentous mistake, but a clear conscience and an intact reputation for integrity are infinitely more valuable than an illicit 'A+.' In general, for all of your academic activities – in the teaching and research laboratory or otherwise – it is critical that you maintain the highest personal standards of academic (and ultimately professional) conduct and be honest and responsible in executing and presenting your work.

Your professor has an ethical obligation to you as well. If you have reasonable concerns about or potential difficulties with any aspect of the laboratory experience, such as allergies to particular chemicals or materials (such as latex gloves) be sure to share those with you teaching assistant or professor. Support for many other types of accommodations in conventional classroom settings and in laboratories is usually available as well through dedicated campus offices that are focused on disability services, for example, within the college community. They will typically work with faculty members and others as necessary to reduce barriers as much as possible to allow all students to succeed.

NOTES

1   You will find that scientists will disagree even on what the scientific method is. That "There is indeed no such thing as 'the' scientific method" is a claim from one commentator: *The Limits of Science*, Peter Medawar, Oxford University Press, 1984 (see pp. 50–52). The presumption that the 'scientific method' is a road map to sure success, "a kind of calculus of discovery" is a presumption challenged in that short book.

2   A technical treatment of the development and nature of science is offered by *The Natures of Science*, Neville McMorris, Fairleigh Dickinson Univ Press, 1989.

3   Even if constructing and stating a hypothesis is not always necessary, this article, written by ecologists but of broader interest, argues for its usefulness in many cases, and even the value of considering multiple hypotheses: Betts, M. G.; Hadley, A. S.; Frey, D.W., et al. When Are Hypotheses Useful in Ecology and Evolution? *Ecol. Evol.* **2021**, *11*, 5762–5776.

4   A mountain in Antarctica has been named 'Mount Hypothesis' as a tribute to the historic role of the hypothesis in the scientific method: *The Composite Gazetteer of Antarctica of the Scientific Committee on Antarctic Research* – Place ID: 20160; Name ID: 139232.

5   These are errors that arise because of the position of the reader relative to the instrument. To read the volume of a liquid in a measuring cylinder or burette, for example, you should have your eyes level with the center of the meniscus. That usually means to the bottom of the concave meniscus. In cases where the meniscus is convex (e.g., liquid mercury) such that the center of the meniscus is above the liquid-container interface, the reading is taken from the top of the meniscus. Similarly, we should read analog clocks face on, not from the left or the right, to minimize errors.

CHAPTER SEVEN

Spreading the Word

Communication skills are as critical in the sciences as they are in many other spheres of life. Scientific communication has its unique features, but a basic requirement for success as a student in any discipline is the ability to receive, process, and share information effectively. In the sciences, communication skills are called upon in simply talking to your professors during office hours, completing and submitting problem sets or exam papers, participating in classroom discussions, defending honors theses, and so on.

So, what does communication have to do with problem-solving? You are probably familiar with this philosophical question on the role of perception: "If a tree falls in the forest, and nobody is there to hear it, does it make a sound?" For problem-solving, a related question might be this: "If you solve a scientific problem (or answer any significant question), and nobody else knows about it, has it really been solved?" Many scientists would probably say "No" to that question. New scientific knowledge should be made known. The word should be spread, and well beyond the philosophical position, it is generally understood that scientists have an ethical obligation to share news of discoveries from which others could benefit.

The typical ways of sharing such news in the sciences include published academic journal articles, posters, and oral conference presentations. Once a problem is identified, investigated, and solved (or new insights are gained into the nature of the problem, or new tools developed to tackle it, etc.), the news of your success or lessons learned from instructive failures should be communicated to the relevant audience, be it the professor for your course, or readers of a professional research

DOI: 10.1201/9781003263340-7

journal. We make some brief remarks here on some of the common modes in which such news is still regularly shared.

Preparing Papers

Reading good books by good writers and engaging with your friends in mutual, honest, and helpful criticism of each other's writing can strengthen your writing. When you are preparing a paper for a specific assignment, make sure that you are clear on the topic and the requirements for the paper. What are the spacing and margin requirements? How many pages are allowed? Do you need a dedicated citation section at the end or should you use footnotes? And so on. If in doubt, always assume that you should include references in some way. This will help you to avoid, too, any hint of plagiarism. Other questions: How should references and notes be formatted? Are there limitations and style preferences for figure and table captions? Is an abstract required? For papers in college, anything that is not specifically required by your professor can be decided at your reasonable discretion in line with norms from previous papers that you have written for that or similar types of courses, but you can also double-check with your professor. If you decide on your own to do something odd or exceptional – like putting only the word 'simple' (in red, in the center) on the second page of what should be a three-page essay on the philosophy of simplicity – you should know, at least, that you are rolling the dice on your grade for that paper. Even with high levels of creativity, it is usually important that you stay centered on the topic at hand and honor the goals of the assignment. All of the style and innovation that one can infuse into a paper or project will not help usually if you fail to meaningfully address the actual learning goals, themes, or topic of the assignment.

Different types of essays will require somewhat different styles. An argument against sending animals on test flights to Mars, for example, would call for a different tone (and a different structure too, possibly), from that of an essay on the impact of the industrial revolution on family life in England. There are many resources available online and through your college's writing center on writing well – on the use of language, basics of argumentation, the mechanics of effective essays, and so forth – and you should feel free to draw guidance from those sources. As for materials relevant to the actual topic under consideration, you are free, too, to draw, within the parameters of the assignment, on hard copy or electronic resources, such as dictionaries, thesauri, reference books, news magazines, and relevant research publications, that are available to you as you frame and write your essay, but, again, be sure to avoid plagiarism. Intentional plagiarism insults your intelligence, impugns your character, and denies your mind the privilege of finding and articulating an innovative perspective. Excessive generosity in your references is an offense that your readers and you ultimately will find to be much more forgivable than any organized effort to claim the work of others as your own. Your professor and any college librarian will be able to help you to locate and use web-based and local applications dedicated to retrieving and assembling relevant reference data (by topical (keyword) or author searches, for example). Those search results will typically includes citations for books, journal articles, and so on, and you may even be able to land actual copies of publicly available documents in some cases. If citations of interest are behind paywalls, you might have to work with one of the college librarians to try to secure a copy.

Beyond the acknowledgment of sources on which you have drawn in writing your paper, keep in mind that your list of references also provides others with some insight into the kinds of sources that have influenced your thinking on the topic. References can also provide readers who are also very interested in the topic with new sources that they might consider in the course of their own work.

If you finish a near-final draft of your paper with some time to spare, ask one of your peers to read it and offer any helpful feedback that they might have. An alternative, if that is not possible, is to step away from the text for a day or more, if you can afford it, and take another look later on with fresher eyes. You may find new things that you would like to eliminate, modify, or amplify that will cause you to feel better about the final product. Two books on writing that I have found to be insightful are *The Elements of Style*,[1] and *On Writing Well*.[2]

Writing Abstracts

Many college papers that you will write will be conventional essays followed perhaps by bibliographies or references. Many academic papers in science and other disciplines will require an abstract, too, that will precede the full essay or scientific report. The abstract is an 'introductory summary' of your paper that has a word limit in many science journals of about 200 to 250 words. It's 'introductory' because it comes at the beginning of the paper, is usually the first part of a paper that someone will read, and it is your chance to welcome the reader into an intellectual engagement on a topic of interest to you. It's a 'summary' because it gives the reader a brief, clear, and compelling picture of what your essay or scientific report is about. Some readers will decide whether to continue reading your paper in detail based on what they infer from the abstract about the content and the quality of the paper – including the quality of the writing. So, an abstract can make or break your paper. If it is too long, it suggests that you do not know what is important in your paper, have severe difficulties with brevity, or do not respect your readers' time enough to do the work of writing a strong and short abstract.

To prepare your abstract, you might consider the following: what the work is about – the motivations for the work – major results – major conclusions – broader implications – sources of error, and limitations to or key assumptions made in the study.

An Example: Suppose you went to the Sudan to study desert insects, specifically 'Q Bugs.'[3] An abstract for your report might read as follows:

Abstract: The insects of the Sudan are remarkable in many ways, with certain features that are unique to the region. We have investigated the lengths of the hind legs of the Q Bug and the impact of leg lengths on life span in the upper desert region. There has been some speculation that leg length correlates with life span for other critters in Australia, but little is known about the Q species in North Africa. We find that there is no correlation whatsoever. The average life span for all of the 500 Q Bugs that we studied over six months is 42.5 ± 0.3 days, with averages of 42.4 ± 0.1 days for the 100 bugs with the shortest legs and 42.6 ± 0.1 days for the 100 bugs with the longest legs. The reason for the differences in the correlations observed in the Sudan, as reported herein, and those reported previously for similar species in Australia remains unclear. Hypotheses for future investigations into the differences observed are outlined near the end of this paper. [174 Words]

This fictional abstract includes five sections in under 200 words. So even more could be said if one wished, but the temptation to say everything should be resisted. In this case, the actual lengths of the bug legs, and how they were measured, are excluded. Should that information have been included? Perhaps. But those are the kinds of decisions that you will have to make as you work to achieve meaningful and engaging messaging, clarity, and brevity in an abstract – and keep in mind that details excluded from the abstract will still be available to readers in the body of the paper. The sections in this mock abstract are (i) an engaging introduction to the question, (ii) a comment on what this paper does, (iii) a relevant reference to the state of the field, (iv) some key results of the paper, and (v) some positioning of the current results in context plus hints of possible future directions. Professors will differ somewhat on exactly what and how much they would like you to include in an abstract. Some disciplines might even have rather detailed or rigid requirements, but brevity and clarity, in this regard, are always virtues. A conversational style tends to be more compelling for readers, even in the sciences. Also, even as some academics still favor the third-person and other traditional modes for writing research papers, writing in the first person, as we did in the abstract above, has its benefits. Scientific reports written in the first person can engage and connect quite directly and transparently with readers, admitting implicitly that the authors, like the readers, are just people interested in a scientific question. In the first or third person, however, the goal remains the same for the abstract and for the entire paper: to write honestly, logically, and as accessibly as possible (for the widest audience possible) given the topic, the results, and the implications of your work.

Preparing Posters

Some of your college courses may require you to prepare a poster on the outcomes of an individual or group project, and undergraduate research students often prepare reports on their work in poster formats as well for special research symposia. Such presentations often precede any publication of their results as peer-reviewed papers in academic journals. So, poster sessions tend to be the first opportunities that successful research students have to share their work with the wider academic community and the public.

Keys to preparing a good poster and presenting it well: gaining a deep understanding of your results, designing an appealing and readable poster, and having a coherent message that you can deliver clearly and succinctly. Unless your research group has a rigid format for undergraduate posters, you will have some real autonomy over the appearance of your poster.

I include here two model poster layouts (Figure 7.1). The poster design at the top in that figure is more traditional. The second one, which I composed just for this purpose, may be too scandalous for some and quite appealing for others. The options are infinite; the choice is often yours. Whatever your preference in style, however, convey your information clearly and as transparently as possible. Brevity is important, and clarity is more important. Keep in mind, too, that *style is content –* the 'Summary' sections in the model posters, for example, are different in length, because the second format actually affords less space than the first.

Your figures, such as graphs and illustrations, should be clear, including any text that is in them. The axes of your graphs

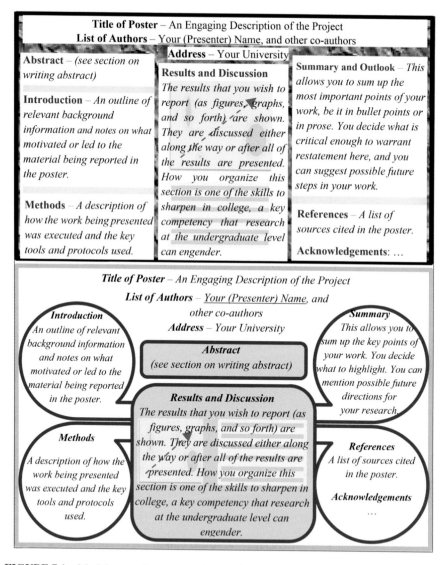

FIGURE 7.1 Model poster formats: a rather traditional format at the top and an alternative at the bottom.

should be properly labelled with the names of the quantities plotted and the corresponding units, for example, Volume/m³. In many cases in the sciences, captions are placed below figures but above tables. Be sure to check what the conventions are for your specific discipline. Although the illustrations in Figure 7.1 contain a lot of text, the words are descriptions of what you might include, such as figures, and tables. You should avoid having too much text on your poster. A typical passerby at a poster session will not stop and read it in detail. You will be doing much of the literal talking, but you can have the poster help you in that process by showing graphs of your data, mechanisms for reactions, diagrams of your experimental set up, and so on – pictures, after all, are reputed to be worth a thousand words. They will help you do the talking. Some explanations in words on the poster will help for sure. The abstract, introduction, and methods sections (see Figure 7.1) are often mostly text – an explanation of your motivation and approach can help the casual reader to get a sense for what your work is about – but the combined results and discussion section is a large canvas on which you can show more than you could effectively tell with text alone. The summary section allows you to highlight again the key outcomes from your work.

Research can be costly. Your research mentor may ask you to include on your poster a logo or symbol representing your university or an external funding source for your research such as a government agency or a philanthropic organization. That is a professionally acceptable way to gratefully acknowledge their support. It is also an acceptable way to declare the sources of your funding. If your research on *Health Benefits of Bread* is being sponsored by the company 'Breads-Are-Us Inc.,' the public should be made aware of that sponsorship when you report your results. That kind of transparency is healthy. It keeps you honest, and it allows others to be fully informed in making their own assessments of your work and any potential conflicts of interest.

You and your research advisor get to decide what aspects of your data and analyses will make it onto the poster. You do not have to report everything that you have done, but you have to be willing to engage coherently with curious minds about what you decide to present. In general, the audience will not be out to get you on any error, omission, or technicality as you discuss your work with them during a poster session at a scientific conference. They will want to understand your work and to support your success; and their questions, suggestions, or even respectful criticism, should be taken in that spirit.

Poster presentations are usually given in large rooms with people standing beside their posters, which may be printed and pinned on to presentation boards or displayed in some other way.[4] On your college campus, the people attending the poster session will usually be your peers, faculty, and possibly others who are interested in the discipline or topics related to the focus of your poster. At large external conferences, the audience will include undergraduate and graduate students and scientists working in academia and industry from several different national and even international universities. Whatever the context, those who are not presenting posters will be milling around the room gazing from a distance at the posters, drawing closer to ask questions if the specific topic or appearance of the poster entices them. Who, and how many people will come to your poster, and how many questions they will ask, are impossible to predict. You will get a lot of attention at some poster sessions and less at others – especially at large conferences, it will depend on where your poster is located and who happens to float over to your section of the room. Poster sessions might last an hour or a bit longer. I usually encourage students to spend most of that time beside their poster responding to visitors, but I also encourage them to take a few minutes of that time to step away from their poster and mill around as well. That brief excursion enables them to see what others are working on and to hear some of the stories of other student researchers. I think that can be very useful for student exposure and development, if only by helping the student to see that some of the questions, joys, and challenges that they have met in tackling research questions are shared by peers and professors elsewhere. I'm not sure that all research mentors subscribe to that idea; check in advance with your advisor or mentor to see if it would be fine for you to take a few minutes to see what other presenters are working on in their research groups. Some conferences hold multiple poster sessions, however, in order to accommodate large numbers of posters, so students are free to peruse during other poster sessions before or after the time slot to which they are assigned.

Presenting a poster anywhere – especially at a large conference, to people you do not know – sounds ominous. But it will likely not be as scary at all as it sounds. Here, as in so many other areas of life, two elements that will foster success are preparation (leading up to the presentation), and, as best as you can achieve it, relaxation – especially during the presentation.

Prepare for your poster presentation by practicing to deliver a very brief description of what's on the poster (using a friend outside of your discipline as the audience, if possible). The target should be to introduce yourself and to summarize what you did, why, and the key results, all in a minute or so. (*E.g.: Hi. I am Jane Brown from the Stewart group at Great University. My research focuses on . . . We have found that . . . as we show here. Our next step will be to confirm that . . . I'd be happy to share more of the details with you or answer any question that you might have.*) After that, you can hold a more detailed discussion with the person at your poster based on their questions or feedback. Brief summaries, where you have a short time to deliver a clear and cogent message of what you are up to (in this case, what your poster is about), are sometimes called 'elevator speeches.' That term (also 'elevator pitch') is used in business contexts where stereotypically you might only have the time between a few floors in an elevator to say who you are and what you are working on or proposing. The ability to send an important message in a few sentences is a life skill that poster presentations can help you to gain and strengthen.

Presenting posters, like other forms of expression – be they written papers, seminar talks, stand-up comedy, or acting – will usually improve with practice and experience. So, do not be hard on yourself if you think at the end of the poster session that you mis-answered a question or fell short in some way. Every poster session is a chance to share what you know, to learn from others, and to grow as a thinker, presenter, and as a person.

Preparing Talks

An oral presentation supported by slides or demonstrations is an excellent way to report the results of a study or research project, or to share a new idea. The information that you will be presenting may have been generated by others and you are simply reporting it, or it may be the result of data mining or original research that you completed alone or as part of a team. One can give a talk on almost anything, but the key to a successful presentation is to be persuaded that the material that you are presenting is interesting and important (to you at least, and hopefully to the audience as well). If you are persuaded that the issues to be discussed in your talk are important, you should reflect that fact in your preparation to deliver the talk (making time to envision and develop the talk, assembling materials or samples that you might want to show to the audience, and practicing the delivery of the talk). The quality of your presentation – for example, voice volume and control, eye contact with the audience, honoring the allotted time, factual and spelling accuracy, the legibility of text on your slides, suitability to the knowledge level and interest of the audience, the clarity of your message, and so on – plus your enthusiasm in delivering the talk, and your willingness to answer questions based on the talk, will be influenced greatly by the quality of your preparation.

Worrying about some of the aspects that I just mentioned (understanding your results, organizing your data, clarity, brevity, and so on) might cause you to think that giving a talk to a group is a terrifying experience, but it does not have to be. Confidence and success come from practice; not just practicing before your first talk, but by entering the arena and learning from each talk that you will both see and give. Many presenters find that a lot of the anxiety about presenting actually fades away once they start talking at the beginning of the presentation.

The audience knows that you are not omniscient – you will not need to know everything about all topics before you can prepare and deliver your talk on the aspect of the subject that you have been studying. Additionally, it is crucial that you be transparent and honest as a presenter: be willing to state as clearly as possible (during your talk or in response to questions from the audience) what the successes, limitations, and even shortcomings and failures of your study were. If a typographical error escaped your careful preparation, and you notice it while you are giving the talk, do not be flustered. Apologize for it, or just note it and move on if it is a trivial oversight, or explain what should actually be there on the slide instead, if that information is critical to understanding the talk. If you do not know the answer to a question, say so. Say that you are unsure, if that is the case, or give the best answer you can if you can manage to make a reasonable deduction based on everything else that you know about the subject. It's OK to proffer a hypothesis on the spot if you are comfortable doing so or to indicate to the questioner that you will have to think a bit more about the issue or run more experiments to find an answer to the question. If you make such promises, though, be sure to keep them. That questioner might show up at your presentation again next year – a sign perhaps that they are really curious about or truly interested in your work.

One way to get better at giving talks is to listen to good talks, even talks that are outside of your immediate area of academic interest. There are many talks available for viewing online and you will see copious announcements for free lectures on your college campus. Some people will give their talks to a small group of friends or close colleagues just to see how 'friendly' audiences react to the material that they plan to present at the official event.

Your slides should be well-ordered, tailored for the time limit, and numbered. Slide-1 typically shows the title of the talk, your name, the date, the name of your course (if it's a class presentation) or the name of your organization (your college or a group you are representing), the name of the event (if it's a conference talk, for example), and so on. If you are reporting on a project that you worked on with others, you can present their names on this title slide (especially if they will be participating in the presentation with you) or you can highlight their contributions later on, on an acknowledgment slide. You can follow the title slide with a set of introductory slides that will include some important background information to help the audience understand the historic significance and current relevance of the topic.

As a former mentor advised me, "No Outlines!" Do not include a slide that forces your audience to watch you read to them a detailed list of what you will be telling them. The title and the introduction are enough. Let the details unfold as the talk unfolds. That said, there is no divine rule here; include your outline – after the title slide perhaps, before the introductory section – if you think, for instance, that it will comfort the audience to know what is coming, or that mentioning upfront an amazing revelation coming near the end of the talk ("A Solution to the Moving Sofa Problem!") will keep the audience transfixed until the end of your talk.

A description of the methods used in the study and the main body of the talk (the results and discussion slides) typically follow the introduction. The point made earlier about the value of figures and tables in posters applies here as well. Use clearly labelled and appropriately captioned figures and tables, well-developed or selected videos and so on to help with minimizing the reliance on the text on your slides. It might help if you use a simple and consistent layout for presenting your data, but you will have many good options for how to organize and share your results. Will you split each slide in the results and discussion section in half, and show a figure or table on the left and supporting text on the right in each case? Will you show only an illustration of your data on the full slide and rely on your spoken words to clarify the message of the slide, or will you do something else – use a combination of formats even? Whatever choice you make in displaying and describing your results, the selected format should serve to enhance rather than distract or detract from your mission and your message. The penultimate content slide is usually a summary list of the key results from your talk, the key implications of those results, and possible future directions. The last slide with any technical content might be your list of references or a short bibliography on the topic. If you include relevant references on individual slides as the talk progresses – an approach that I find to be more helpful for audiences – then a dedicated reference slide is not needed at all. The final slide is usually your acknowledgments. Thank collaborators (your research professor and others), any funders, and others who supported you and your research efforts.

A useful practice is to include a few potentially helpful slides after the last official slide of the talk. Think about what people might want to know beyond what you will have time to talk about. It may be an extra graph or diagram. If you were studying cave art, and you had 1000 pictures of cave drawings, but only showed 46 of them in the body of your talk on 'The Primacy of Cats in Ancient Cave Art,' include perhaps – as extras on slides after the official final slide of your talk – some additional pictures of an even wider diversity of the types of cave art that your research uncovered. You may not show them at all, but someone in the audience, who happens to be studying ancient drawings of sea creatures, might ask about fish in cave art during a discussion session, and one of those extra slides, though irrelevant to the feline focus of the talk, might come in handy as you respond to that question.

How many slides should you plan on presenting? Enough to convey the message that you plan to deliver within the time that you are allowed, plus or minus one. Plus one, if you are the kind of person who goes too slow when you practice, but speeds up when you are before an unfamiliar audience; minus one, if you do the opposite. In any case, a general rule of thumb is that one slide per minute is usually too much. A fifteen-minute scientific talk should probably not have fifteen slides; we often start talks by introducing ourselves or the title of our talk, and that alone might be one minute or more gone forever – leaving, at most, one minute per slide for the rest of the talk, and no time at the end for discussion. If you will be delivering the talk in a setting where audience members are free to ask questions at any point during the presentation, time will be needed for that too. That will be difficult to plan for, but the best way to see if you have too many or too few slides for the time allocated is to practice in as realistic a way as possible. Short talks warrant a lot of practice since they allow you little room for error or diversions. For a talk of any length, however, practice will make you more comfortable with the material and will hopefully relieve anxieties enough that you start to look forward to finally delivering the grand product of your research and rehearsals.

With all of this advice on form and style, though, be sure to build a talk that you like and will enjoy presenting. Again, a great way to learn about talks, beyond giving many talks yourself and embracing those experiences, is to watch others present and reflect on their approaches. That can give you a sense of how much is too much on a slide, and what colors, fonts, and background layouts work well for slides, for example. I have learned and continue to learn a lot from seeing others present. Here, again, there are helpful resources to which your university's speech center or library might have links or access. Many great talks on diverse subjects are also available online for free.

The mood and enthusiasm of the presenter (of a poster or a talk) can strongly influence the reaction of the audience to the presentation. An unengaging talk on *exploding stars* can be enervating for an audience while an engaging talk on *silence* can be energizing. Starting your preparations early, practicing the talk, and getting feedback from others can help with any upcoming presentation, and – in the long run – the experience itself of giving more talks will make a positive difference. After all of that, if you find joy in delivering your talk, it's more likely, I think, that the audience will find joy in receiving it.

NOTES

1 *The Elements of Style* William Strunk Jr., E. B. White, Pearson; 4th Edition, 1999.
2 *On Writing Well: The Classic Guide to Writing Nonfiction*, William Zinsser, Harper Perennial; 30th Anniversary. Edition. 2016.
3 This is a fictional bug, to be sure. So too is the abstract. It's constructed to illustrate the basic form and general elements of an abstract. What you include in your abstract will be dictated ultimately by the nature of your study and the key message that you wish to deliver with your report.
4 Depending on the nature of the presentation and the context, the poster might even be shown on a screen – an option for small class presentations, for example.

CHAPTER EIGHT

Persisting against Problems

Mindset and Anxiety about Belonging

If you feel academically unprepared for a course, especially for your first semester of college, talk with your professor before or at the start of the course. In most cases (judging from my experience with anxious undergraduates), you will be underestimating your current knowledge, skills, and abilities. By the end of high school, students usually have or will gain quickly the technical fortitude to master the new material ahead. Keep in mind, as you peruse syllabi for your courses, that you will be expected to understand the material during and by the end of the semester, and not all at once at the start of it. Do not be intimidated or think that you cannot succeed in a class because much of the material looks foreign to you at the outset. If you knew the material already, you would not need to take the course.

As classes get underway, you will be advancing toward mastery of the content, and competencies that you will need in the second half of the semester, for example, will be strengthened or newly acquired during the first half of the semester. In some cases, you might be less prepared indeed, for one reason or another, than your typical peer in the course. And professors are there to help. Make use of faculty office hours, which are times specifically dedicated to meeting with students. Talk to professors about where you see your preparedness for the course, and ask for suggestions on how to efficiently and effectively ensure your success in it. Assume

DOI: 10.1201/9781003263340-8

that your professors have high expectations of you, and allow that to build on your high expectations of yourself. Like office hours, academic skills centers, student success centers, or whatever jazzy name they use on your campus, exist to support your progress. Many institutions will provide some access to tutors or other forms of near-peer support that can help you to transcend what seems in the moment to be unsurmountable hurdles.

For college students, wrestling to fully grasp a new topic or subject are signs of commitment and engagement, not signs that the subject is too difficult for you or that you do not belong. Mastering the subject was likely not a smooth ride for your professor, and many of your peers will also have challenges, but because we all are more aware of and more sensitive to our own struggles, those struggles can cause us to feel isolated, outside the norm. But struggle of a certain form is a part of learning.[1] Remarkably, even demonstrably high achievers, some of whom have established themselves as successful professionals, suffer unjustifiably from "*an internal experience of intellectual phoniness*"[2,3] that has come to be called the impostor phenomenon (or syndrome). Many factors can lead to this feeling – the origins can be quite personal or from the social and cultural context in which we live and work – but for college students (as for qualified professionals in the workplace) such perceptions are typically out of line with your record of accomplishments, and your abilities and promise. Get advice and counseling as needed; whatever you do, you should not allow momentary feelings of phoniness to frustrate your will, impede your academic success, or quench valid personal and professional aspirations.

Most of your professors and peers will acknowledge your abilities and affirm your potential for achieving excellent academic outcomes, but if others focus on perceived deficits that they think you have, do not join them in doing so. You will be able to work strategically (reading, practicing, and so on) on your own, and you can meet with the instructor at intervals as needed to discuss key topics and your progress in the course. But always work from the vantage point that you are using your intellectual strengths to gain new proficiencies *en route* to your desired outcomes. Your established record that got you into college is good evidence that you belong and can succeed there. Academic success is a fruit of an iterative process in which you use what you know to learn what you don't.

Thoughts on Managing the Demands

AVOID OVERCOMMITMENT

"No" is an acceptable response. Be sure that you have time for (and a real interest or stake in) a thing before you commit to it. Many of us have problems saying '*No*' when we should, but, as they say, practice makes perfect. So, work at saying "No" when that's what you know that you should or simply want to say. Feel free to say it as it is: "I cannot make it," "I have another commitment," "I'm not interested," "That's not for me right now," "I am not available – I have things to do!" as the case may be. It may not seem so initially, but students respect their peers who have their priorities in order. In time, your example might even fortify more timid peers to resist pressure and think again as well before sacrificing a high priority for a low one, or for something that's not good for or even interesting to them.

GET A CALENDAR

Electronically or on paper, schedule the things that you need to do, such as attending your classes and fulfilling other academic commitments, studying, watching or playing a game, and so on. You do not have to be rigid when unexpected demands arise, but you should have a plan. Do not live to indulge the next arbitrary whim of a friend, or worse, others with no interest at all in your success.

SLEEP – EAT WELL – EXERCISE

Schedule sleep and times to eat into your routine, and go to the college gym, or play in pick-up games, take a tennis class . . . or be active in some other way. Problem-solving benefits from a clear mind and physical health. The quality and duration of your sleep, the nutritional value and balance of your diet, combined with opportunities to be active and to engage with others, go a long way to foster both mental and physical well-being and success.[4]

NURTURE GOOD FRIENDSHIPS

Friends will come; they tend to arise naturally out of coincidences or shared academic, recreational, or other interests and activities on campus. Your closest friends should be those who encourage you in the right direction and provide support and advice that will make you proud of your choices a decade from now. Good friendships are blessings that you should count as you assess the resources available to you, as you confront problems in college beyond your academic commitments. You should value and nurture fellowship with those who strengthen you as a student and as a person more broadly.

BE GOOD TO OTHERS

You cannot be everyone's best and closest friend (and you should not try to be), but as much as you can, be kind to others with whom you will live, learn, study, and serve in the college community. Here you might consider others who are less well-off, less popular, and less represented than you are, be they fellow students or service staff, for example. An attitude of gratitude is a big plus as you live in a community, even if not in communion, with others.

REMEMBER WHY YOU ARE IN COLLEGE

At its center, your mission is, no doubt, to get the best academic training and personal development possible in preparation for your next academic and professional steps. You are not in college, as some claim, to find yourself. Sure, you will learn more about you and the world, but self-discovery does not require college. A part-time job, travel, quiet reflection, and the regular adversities and joys of life in any context will help with that self-finding as you get older and more mature. The college environment is, however, a unique context in which to grow intellectually and in many other ways in a community with motivated peers and dedicated mentors. You will gain insights and inspiration well beyond what a signed degree will tell when it is won, framed, and positioned on a wall just a few years from now. But remember to articulate your key academic and other developmental goals for college, and keep them at the center. This takes us back to the list that you made early on in the first chapter. Your profile and reputation on the college campus will depend on what you identify at the outset as your academic and personal growth objectives for your undergraduate experience. To make a problem-solving analogy, the opportunities that you will see as a 'blessing' toward the goal of engaging successfully with others and the broader campus community, will depend on what exactly 'successful engagement' means to you in practice.

Be Gracious and Forgiving

This should be applied especially to yourself. Success can often come after many setbacks. Do not be hard on yourself after a disappointing outcome – if a grade did not land where you wanted it to, you made an error of some sort in class or otherwise, or you feel disappointed with one thing or another. Most of the people who seem to always have the answers (professors, teaching assistants, or classmates), have hard days too, even if it does not seem so now. You are not the sum of the grade you got yesterday or even the degree you will earn tomorrow. You will find in past failures some of the greatest wisdom and encouragement that you can give to others – future students included – after you have triumphed with the joys and worked through the challenges of your own experiences. Be gracious and forgiving to yourself, and extend similar courtesies to others.

Celebrate

Find sober and healthy ways to affirm and rejoice in your successes. You made the Dean's list, published your first article in the student newspaper, completed a project, delivered a presentation at a conference, and got the campus job or internship you wanted. Relish the victories. Acknowledge and celebrate them, even in small ways such as making a note of them in your journal, if you have one, or sharing the news with friends. Mark the moments. This applies to completing a demanding academic assignment, winning a class debate, getting into graduate school, and every other reason to rejoice. In a quiet college exam room with other students, a smile and mumbling "Yes! Awesome!" might be enough merriment for crushing a big question during a test. If it's winning a coveted scholarship or a campus student award, you might call your grandmother, share the news, catch up with a few close friends, and celebrate.

There will be points now and again in the problem-solving process when despondence sets in, when a question seems intractable, a paper seems unwritable, and experiments for a laboratory course or undergraduate research project seem unworkable. And you will get beyond those points, but marking current victories can energize you for times like those when you take on your next big challenge.

On to the Next Problem

Life in all its facets presents us with questions to answer, mysteries to unravel, and problems to solve. Some, as it were, will require divine intervention to resolve; others are quite tractable, requiring, for instance, only a better hammer or a stronger nail. The focus here has been on problem-solving in the college classroom. For aspiring and new college students, your goal for academic problem-solving should be to draw on and advance your knowledge, critical thinking and reasoning skills, your confidence and your priority structure, your willingness to work alone, and your ability to work with others, all in the service of solving the next problem that a college course will present to you. And now – with the intellectual fortitude you have already developed, your teachable spirit, your courage to enter the arena, and your enthusiasm to get started – off you go!

NOTES

1 *Mathematics for Human Flourishing*, Su, F. (with reflections by Christopher Jackson), Yale University Press, 2021. See chapter 8: struggle.
2 Clance, P. R.; Imes, S. A. The Impostor Phenomenon in High Achieving Women: Dynamics and Therapeutic Intervention *Psychother: Theory. Res. Pract.* **1978**, *15*, 241–247.
3 Langford, J.; Clance, P. R. The Impostor Phenomenon: Recent Research Findings Regarding Dynamics, Personality and Family Patterns and their Implications for Treatment *Psychother: Theory. Res. Pract. Train.* **1993**, *30*, 495–501.
4 *Food, Sleep, Exercise: Why You Seriously Need All 3 to Be Successful* Power, R. Success Magazine. July 19, 2017: www.success.com/food-sleep-exercise-why-you-seriously-need-all-3-to-be-successful/ Last accessed February 2, 2022.

Appendix I

Additional Notes on Matrices and Matrix Algebra

The Identity Matrix

The matrix equivalent of the number 1, the identity matrix, I, must have the same number of rows and columns, with '1' at all points along the main diagonal and '0' everywhere else. For a matrix with two columns and two rows (a 2×2 matrix), for instance, the identity matrix, I, is:

$$I = \begin{pmatrix} 1 & 0 \\ 0 & 1 \end{pmatrix}.$$

We can confirm as follows that this matrix has the same effect as unity in basic multiplication. This example will help us as well to see more clearly how matrix multiplication works.

$$\begin{pmatrix} 1 & 0 \\ 0 & 1 \end{pmatrix} \begin{pmatrix} 3 & 4 \\ 1 & -5 \end{pmatrix} = \begin{pmatrix} 3 & 4 \\ 1 & -5 \end{pmatrix}.$$

that is, $\begin{pmatrix} 1 \times 3 + 0 \times 1 & 1 \times 4 + 0 \times -5 \\ 0 \times 3 + 1 \times 1 & 0 \times 4 + 1 \times -5 \end{pmatrix} = \begin{pmatrix} 3 & 4 \\ 1 & -5 \end{pmatrix}.$

The fact that the I matrix exists implies that we should be able to write a new matrix M^{-1} – in the same way that for any number q we know $1 = q \cdot q^{-1}$ – such that, in matrix terms, $I = M \cdot M^{-1}$. So, for the case in question:

$$\begin{pmatrix} 1 & 0 \\ 0 & 1 \end{pmatrix} = \begin{pmatrix} 3 & 4 \\ 1 & -5 \end{pmatrix} \cdot M^{-1}$$

where M^{-1} is the inverse of the matrix $M = \begin{pmatrix} 3 & 4 \\ 1 & -5 \end{pmatrix}.$

The Inverse of a Matrix

We will show procedurally how to find M^{-1} for a 2×2 matrix, but without any justification for the process. For a matrix $\begin{pmatrix} a & b \\ c & d \end{pmatrix}$ the inverse may be found by dividing what we call the adjoint of the matrix, Adj(M), by a value called the determinant of the matrix, $|M|$. That is:

$$M^{-1} = \frac{1}{|M|} \cdot \text{Adj}(M)$$

The determinant, $|M|$ of a 2×2 matrix M is the number obtained by multiplying the terms that are diagonal to each other, that is, '$a \cdot d$' and '$c \cdot b$,' and subtracting the latter from the former: $|M| = a \cdot d - c \cdot b$.

The adjoint Adj(M) of a 2×2 matrix is obtained by interchanging the terms on the main diagonal, that is, swapping a and d, and multiplying the terms on the other diagonal by -1. So from $\begin{pmatrix} a & b \\ c & d \end{pmatrix}$ we would obtain $\begin{pmatrix} d & -b \\ -c & a \end{pmatrix}$.

Consequently, the inverse of the matrix, $M^{-1} = \frac{1}{|M|} \cdot \text{Adj}(M)$ is:

$$M^{-1} = (a \cdot d - c \cdot b)^{-1} \cdot \begin{pmatrix} d & -b \\ -c & a \end{pmatrix}$$

So, for the matrix that we introduced above, $\begin{pmatrix} 3 & 4 \\ 1 & -5 \end{pmatrix}$,

Where $|M| = a \cdot d - c \cdot b = (3 \times (-5) - 1 \times 4) = -19$ and

$$\text{Adj}(M) = \begin{pmatrix} -5 & -4 \\ -1 & 3 \end{pmatrix}$$

We find that:

$$M^{-1} = \begin{pmatrix} \frac{1}{-19} \end{pmatrix} \begin{pmatrix} -5 & -4 \\ -1 & 3 \end{pmatrix} = \begin{pmatrix} 5/19 & 4/19 \\ 1/19 & (-3)/19 \end{pmatrix}.$$

$$M^{-1} = \begin{pmatrix} 5/19 & 4/19 \\ 1/19 & (-3)/19 \end{pmatrix}: \text{The inverse matrix for our problem.}$$

To confirm that the latter matrix really is the inverse of $\begin{pmatrix} 3 & 4 \\ 1 & -5 \end{pmatrix}$, let's test it directly.

$$\begin{pmatrix} 3 & 4 \\ 1 & -5 \end{pmatrix} \cdot \begin{pmatrix} \dfrac{5}{19} & \dfrac{4}{19} \\ \dfrac{1}{19} & -\dfrac{3}{19} \end{pmatrix} = M \cdot M^{-1}$$

$$= \begin{pmatrix} \dfrac{(3\times5+1\times4)}{19} & \dfrac{(3\times4+4\times-3)}{19} \\ \dfrac{(1\times5+-5\times1)}{19} & \dfrac{1\times4+-3\times-5}{19} \end{pmatrix}$$

$$= \begin{pmatrix} 19/19 & 0/19 \\ 0/19 & 19/19 \end{pmatrix} = \begin{pmatrix} 1 & 0 \\ 0 & 1 \end{pmatrix}.$$

SOLUTIONS

So, we did it! We found the inverse. But our goal was to solve for f and g; how does all of this interesting analysis help?

Well, we had the equation $\begin{pmatrix} 3 & 4 \\ 1 & -5 \end{pmatrix}\begin{pmatrix} f \\ g \end{pmatrix} = \begin{pmatrix} 18 \\ -13 \end{pmatrix}$ in our earlier work to solve the simultaneous equations, $3f + 4g = 18$ and $f - 5g = -13$.

And, using the definition $M = \begin{pmatrix} 3 & 4 \\ 1 & -5 \end{pmatrix}$, we can rewrite the

matrix equation as $M\begin{pmatrix} f \\ g \end{pmatrix} = \begin{pmatrix} 18 \\ -13 \end{pmatrix}$ and multiplying both sides

by the inverse matrix, M^{-1}, gives

$$M^{-1} \cdot M\begin{pmatrix} f \\ g \end{pmatrix} = M^{-1} \cdot \begin{pmatrix} 18 \\ -13 \end{pmatrix} \Rightarrow \begin{pmatrix} 1 & 0 \\ 0 & 1 \end{pmatrix}\begin{pmatrix} f \\ g \end{pmatrix} = M^{-1}\begin{pmatrix} 18 \\ -13 \end{pmatrix}$$

That is, $\begin{pmatrix} f \\ g \end{pmatrix} = M^{-1}\begin{pmatrix} 18 \\ -13 \end{pmatrix}$

So we need the inverse of $\begin{pmatrix} 3 & 4 \\ 1 & -5 \end{pmatrix}$, which is $\begin{pmatrix} 5/19 & 4/19 \\ 1/19 & -3/19 \end{pmatrix}$

$$\therefore \begin{pmatrix} f \\ g \end{pmatrix} = M^{-1}\begin{pmatrix} 18 \\ -13 \end{pmatrix} = \begin{pmatrix} 5/19 & 4/19 \\ 1/19 & -3/19 \end{pmatrix}\begin{pmatrix} 18 \\ -13 \end{pmatrix} = \begin{pmatrix} \dfrac{(5\times18+4\times-13)}{19} \\ \dfrac{(1\times18+-3\times-13)}{19} \end{pmatrix}$$

$$= \begin{pmatrix} 38/19 \\ 57/19 \end{pmatrix} = \begin{pmatrix} 2 \\ 3 \end{pmatrix}.$$

$$\text{So,} \quad \begin{pmatrix} f \\ g \end{pmatrix} = \begin{pmatrix} 2 \\ 3 \end{pmatrix},$$

which means that, as we expect from our earlier work on solving simultaneous equations, the solutions are $f = 2$, and $g = 3$.

You can decide if this approach or the other, directly applying Cramer's rule as we did in the main text, is preferred for solving this particular problem.

Appendix II

Thinking about Vectors: Basic Notes

Vectors provide us with a way to talk about quantities such as displacement (movement in a specific direction) and force that have both magnitude and direction. The rules that we apply for matrices are even easier to apply for vectors and we will consider some of the most basic rules that are applied across the natural sciences in using vectors to solve problems.

Giving directions from one location to another is an important responsibility. Tales are told of strangers left bewildered by less than ideal directions received from enthusiastic locals.

> "Go down the road, turn left at the shop, but not the first shop, the second shop with the blue roof, then go farther down the road to the old oak tree with the big limb hanging down, close to where the lady has her beauty supply store, and the auto shop is two buildings down from there on the same road. It's off the road a bit and a little hard to see while driving, but if you reach the school with the pretty bird drawn on the red gate you have gone too far."

A more rigorous way of giving directions – though inconvenient for a person called on suddenly to direct a driver with car troubles to an auto mechanic – are vectors.

The convenient origin or starting point in representing vectors on a graph is the unique position $x = 0$, $y = 0$, and $z = 0$. That is the point where the x-, y-, and z-axes intersect in Figure AII.1.

In three dimensions, these three unique directions, the x, y, and z directions, are all at right angles (or, as we say sometimes, are each *orthogonal*) to each other, and the point where they intersect, (0, 0, 0), is called the *origin*. One important thing to notice is that, starting from the origin, we can travel as far as we want along the x-axis and never go away from $y = 0$ and $z = 0$, and similarly for the other two directions. So, if we

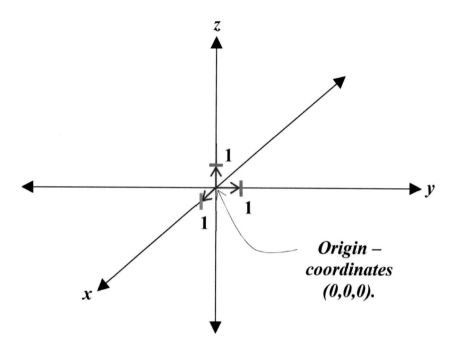

FIGURE AII.1   Model of x, y, and z axes, with the origin at (0, 0, 0) identified. Three unit vectors are depicted for each orthogonal direction.

are at some point a on the x-axis, the coordinates of that position will be $(a, 0, 0)$ and similarly $(0, b, 0)$ for any point b on the y-axis, and $(0, 0, c)$ if we are lounging somewhere on the z-axis.

In any chosen unit of measurement, the vectors $\begin{pmatrix} 1 \\ 0 \\ 0 \end{pmatrix}$, $\begin{pmatrix} 0 \\ 1 \\ 0 \end{pmatrix}$, and $\begin{pmatrix} 0 \\ 0 \\ 1 \end{pmatrix}$ represent a unit displacement in the x, y, and z direc-

tion (i.e., a shift from the origin $(0, 0, 0)$ to the point $(1, 0, 0)$, $(0, 1, 0)$, or $(0, 0, 1))$, respectively. Those nifty looking vectors, which describe the displacement from the origin to a unit distance along the respective axes, are called unit vectors (more accurately, unit vectors in the x, y, and z directions). They are quite useful because they allow us to represent any displacement along the respective axes as products of a constant (a, b, or c) and the relevant unit vector. For instance, if we travel distances a, b, and c in the x, y, and z directions, respectively, we can represent those displacements as follows:

$$a \cdot \begin{pmatrix} 1 \\ 0 \\ 0 \end{pmatrix} = \begin{pmatrix} a \\ 0 \\ 0 \end{pmatrix}, \; b \cdot \begin{pmatrix} 0 \\ 1 \\ 0 \end{pmatrix} = \begin{pmatrix} 0 \\ b \\ 0 \end{pmatrix} \text{ and } c \cdot \begin{pmatrix} 0 \\ 0 \\ 1 \end{pmatrix} = \begin{pmatrix} 0 \\ 0 \\ c \end{pmatrix}.$$

Moreover, we can represent any vector in three-dimensional space (that is any vector with x, y, z components only) as a sum of products of these three unit vectors.

That is, we can rewrite any vector $\begin{pmatrix} a \\ b \\ c \end{pmatrix}$ as the sum:

$$\begin{pmatrix} a \\ b \\ c \end{pmatrix} = a \begin{pmatrix} 1 \\ 0 \\ 0 \end{pmatrix} + b \begin{pmatrix} 0 \\ 1 \\ 0 \end{pmatrix} + c \begin{pmatrix} 0 \\ 0 \\ 1 \end{pmatrix}.$$ (A *linear combination* of the unit vectors)

It matters not what the values of a, b, and c are in the equation.

TERMINOLOGY – BASIS

The set of three unit vectors for three-dimensions is called a *basis* or *basis set* for the three-dimensional vector space because of certain characteristics that the set of unit vectors possess:

(i) we cannot generate any one of the three vectors from a sum of any multiple of the other two. For example, we cannot get to $\begin{pmatrix} 1 \\ 0 \\ 0 \end{pmatrix}$ by adding $\begin{pmatrix} 0 \\ 1 \\ 0 \end{pmatrix}$ and $\begin{pmatrix} 0 \\ 0 \\ 1 \end{pmatrix}$, or by any other so-called *linear combination* of those two vectors: i.e., $a \begin{pmatrix} 0 \\ 1 \\ 0 \end{pmatrix} + b \begin{pmatrix} 0 \\ 0 \\ 1 \end{pmatrix} \neq \begin{pmatrix} 1 \\ 0 \\ 0 \end{pmatrix}$ for any choice of a and b.

(ii) as we just showed, a sum of the three vectors, each multiplied by constants as necessary (a linear combination of the three unit vectors), is all that we need to generate ANY vector imaginable in the three-dimensional space.

We summarize characteristic (i) by saying that the vectors are *linearly independent*, and we summarize characteristic (ii) by saying the unit vectors together *span* the *vector space*. These two criteria are satisfied by any set of vectors that can be said to form a basis. It is actually possible to find other sets of three vectors that are also bases for three-dimensional space, for example, $\begin{pmatrix} 1 \\ 0 \\ 1 \end{pmatrix}$, $\begin{pmatrix} 1 \\ 1 \\ 0 \end{pmatrix}$, and $\begin{pmatrix} 1 \\ 0 \\ 0 \end{pmatrix}$. I'll leave it to you to convince yourself that this set is linearly independent and span three-dimensional space as well – but the set of unit vectors, $\begin{pmatrix} 1 \\ 0 \\ 0 \end{pmatrix}$, $\begin{pmatrix} 0 \\ 1 \\ 0 \end{pmatrix}$, and $\begin{pmatrix} 0 \\ 0 \\ 1 \end{pmatrix}$, is simpler.

You will likely not hear much about 'basis,' 'linear combination,' or 'spanning a vector space' until linear algebra, vectors in physics, or physical chemistry, but you will find that even if these ideas seem strange at first and applications are not obvious they will become useful for solving meaningful problems as you progress in the sciences.

Adding and Subtracting Vectors

Vectors follow the rules of vector algebra, which extrapolate to matrix algebra. Vectors can be added to and subtracted from each other. When we represent vector operations graphically, however, those operations are not as simple as adding two numbers, because we have to take both the magnitude and direction of each vector into consideration.

Consider, for instance, the case in Figure AII.2a, where we have two vectors $\overrightarrow{AB}$ and $\overrightarrow{AC}$ separated by some angle θ. The sum of the two vectors may be written as:

$$\overrightarrow{AB} + \overrightarrow{AC} = \begin{pmatrix} a \\ b \\ c \end{pmatrix} + \begin{pmatrix} a' \\ b' \\ c' \end{pmatrix} = \begin{pmatrix} a+a' \\ b+b' \\ c+c' \end{pmatrix}.$$

And we can show that the resulting vector (what we call the resultant) is $\overrightarrow{AD}$ as illustrated also in Figure AII.2a. You may

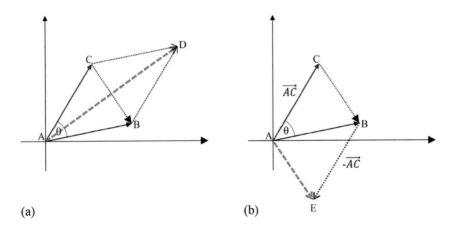

(a)   (b)   E

FIGURE AII.2   Illustrations of (a) the sum and (b) the difference of vectors $\overrightarrow{AB}$ and $\overrightarrow{AC}$. We exclude a third axis here for clarity and we do not specify coordinates for any position on the graphs.

not recognize it immediately, but $\overrightarrow{BD}$ is, in fact, the same magnitude and direction as $\overrightarrow{AC}$. So, traveling from A to D by going first from A to B and then (in accordance with the direction and distance prescribed by $\overrightarrow{AC}$) moving on from B to D is indeed the result of the sum, $\overrightarrow{AB} + \overrightarrow{AC}$.

Subtraction of those two vectors,

$$\overrightarrow{AB} - \overrightarrow{AC} = \overrightarrow{AB} + (-\overrightarrow{AC}) = \begin{pmatrix} a \\ b \\ c \end{pmatrix} - \begin{pmatrix} a' \\ b' \\ c' \end{pmatrix} = \begin{pmatrix} a-a' \\ b-b' \\ c-c' \end{pmatrix},$$

gives, as we can also show, a vector equivalent to $\overrightarrow{CB}$ in Figure AII.2a. Why though is the arrow pointing from C to B, and not in the opposite direction? Well, it become a bit clearer from Figure AII.2b where we strip things down a bit and actually carry out the subtraction $\overrightarrow{AB} - \overrightarrow{AC}$ graphically. The vector $\overrightarrow{AB}$ takes us from point A to point B. The vector '$-\overrightarrow{AC}$' (see Figure AII.2b) would move us to point E such that the net result of the subtraction is $\overrightarrow{AE}$. That illustration (Figure AII.2b) should convince you that $\overrightarrow{CB}$, being parallel to $\overrightarrow{AE}$, provided us with the correct magnitude and direction of the resultant of $\overrightarrow{AB} - \overrightarrow{AC}$.

Of course, if we had values for the angle, θ, and vector lengths, we could draw our parallelograms, as we did in Figure AII.2, or apply the cosine rule, and actually compute magnitudes for resultant vectors for $\overrightarrow{AB} + \overrightarrow{AC}$ and for $\overrightarrow{AB} - \overrightarrow{AC}$.

A question for you: If we did the subtraction in the opposite order, that is, $\vec{AC} - \vec{AB}$, what would the resultant vector look like? You and friends should explore that question and other ideas discussed here. For vector problems, drawing things out, as we have done in this section, will always help. No extra hint will be offered for this question, but you will find that, just as in basic algebra where $2 + 4 = 4 + 2$ but $2 - 4 \neq 4 - 2$, the order of the terms matters for the resultant in vector subtraction.

Other useful operations on vectors, such as the so-called dot and cross products, will become important in various areas across the natural sciences. You are likely to be introduced to them a bit farther along in your academic preparation.

From what we have done so far, we can show that if we have two vectors of the same magnitude pointing in opposite directions (that is where $\theta = 180°$; see Figure AII.3a), the sum of those two vectors will be zero:

$$\vec{AB} + (-\vec{AB}) = \begin{pmatrix} a \\ b \\ c \end{pmatrix} - \begin{pmatrix} a \\ b \\ c \end{pmatrix} = \begin{pmatrix} 0 \\ 0 \\ 0 \end{pmatrix}$$

Two vectors: $\vec{AB}$ and $-\vec{AB}$ Representation of the sum of vectors $\vec{AB}$ and $-\vec{AB}$

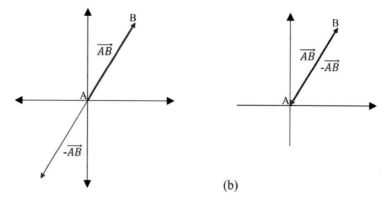

(a)   (b)

FIGURE AII.3 (a) Two vectors equal in magnitude, opposite in direction. (b) Representation of the sum of those two vectors, $\vec{AB}$ and $-\vec{AB}$; since they are equal in magnitude and exactly opposite in direction, we are brought back to where we started and the resultant is zero.

And, if we have two vectors pointing in the same direction and we subtract one from the other, the result will be zero. That is:

$$\vec{AB} - \vec{AB} = \vec{AB} + (-\vec{AB}) = \begin{pmatrix} a \\ b \\ c \end{pmatrix} - \begin{pmatrix} a \\ b \\ c \end{pmatrix} = \begin{pmatrix} 0 \\ 0 \\ 0 \end{pmatrix}$$

and, graphically, the outcome is as shown in Figure AII.3. Since the starting point and end point (after addition) are both at point A, as shown in Figure AII.3b, the resultant vector has a magnitude of zero.

Notice that while we can conveniently represent any vector $\begin{pmatrix} a \\ b \\ c \end{pmatrix}$ as starting at the origin and terminating at some point (a, b, c) (as in Figure AII.3a), the vector simply carries information on magnitude and direction for a quantity. In adding two vectors, the second vector originates where the first one stops (that is point B in Figure AII.3b), not from 0, 0, 0. See also Figure AII.2.

Vectors in the Sciences – A Qualitative Example

Vector ideas are used in various places in many disciplines, in both quantitative and qualitative discussions about displacements, velocity, force, and so on. In introductory chemistry, for instance, vectors may be used in qualitative terms at least to talk about polarity, that is the imbalance or asymmetry in how electrons are distributed along a chemical bond or overall in a large molecule with many bonds.

In a molecule such as hydrogen (H_2), where two identical atoms are bonded to each other, the electrons are distributed equally between the two H atoms. One H atom does not hold the electrons closer to itself than the other H atom (Figure AII.4). For such homonuclear diatomic molecules (of which nitrogen (N_2), oxygen (O_2), and chlorine (Cl_2), are also examples) where two identical atoms are bonded to each other, we say that the molecule is non-polar. The dipole moment of a diatomic molecule is a measure of the polarity of the molecule – that is, the extent of any asymmetry of the electron distribution along the bond between the two atoms. Non-polar molecules have dipole moments of zero. Polar molecules have non-zero dipole moments. Hydrogen fluoride, HF, is an example of a polar diatomic molecule. Fluorine tends to pull the electrons in a bond strongly toward itself (a characteristic that we describe in chemistry by saying that F is very electronegative). In HF, therefore, where the negative electrons are shifted toward F, leaving the positive nucleus of the H atom less shielded, the H side of the molecule becomes locally positive (electron deficient). The accumulation of excess electron density at the F side causes a local net negative charge on that side of the molecule. The molecule is still neutral overall, but the electrons are not distributed evenly, hence the asymmetry in the electrostatic potential across H—F surface in Figure AII.4 compared to H—H. This substantial imbalance in the electron distribution due to the natural (size and electronegativity) differences between H and F means that the HF bond, and hence the HF molecule, is very polar, and we can use a vector to represent that polarity (see Figure AII.4).

For more complicated molecules with more than one bond, it is possible to determine whether the molecule is polar – whether it has a net non-zero dipole moment or not – by thinking about how to sum up the bond dipole moment vectors for all of the bonds in the molecule.

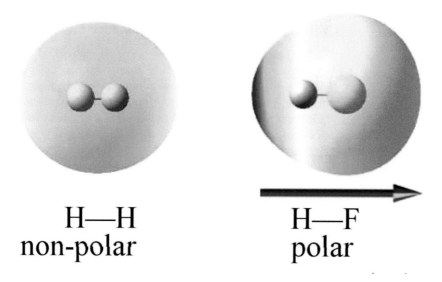

H—H
non-polar

H—F
polar

FIGURE AII.4 Representation of electron distribution and the (non-)polarity of H_2 and HF diatomic molecules. The general color scheme runs from blue (most positive region, stripped of some electron density) to red (most negative region, where the electron density accumulates). The model dipole vector for HF above goes from the more positive H end of the molecule to the negative F end (+ → –).

Beryllium difluoride

Calcium difluoride

FIGURE AII.5 Model structures of BeF$_2$ and CaF$_2$ showing bond dipoles (Not drawn to scale). The bond dipoles are in black and the net dipole for CaF$_2$, which runs through the center of the molecule, is in blue.

Consider, for example, the molecule F—Be—F, which is linear in the gas phase (Figure AII.5). It has two polar bonds, and both of those bond dipoles are equal and pointing in opposite directions, so we expect this molecule overall to have a net (molecular) dipole moment of zero, even though the individual bonds are polar.[1] This situation reminds us of the illustration in Figure AII.3. The molecule F—Ca—F is slightly bent (has a V-shape) in the gas phase, however. So, both bond dipoles will collude (rather than cancel) to confer on CaF$_2$ a permanent dipole moment. Indeed, the fact that CaF$_2$ has a non-zero dipole moment was the decisive piece of evidence that confirmed for scientists that (although both BeF$_2$ and CaF$_2$ might be expected to have the same basic structure in the gas phase, since Be and Ca are in the same group in the periodic table) unlike BeF$_2$, CaF$_2$ is bent.[2]

Vector Ideas in Introductory Chemistry

Take a look at the molecules pictured in Figure AII.6. Some are polar and others are not. Whether a molecule is polar or not can be critical for many experimental investigations and applications of molecules in chemistry, physics, and biology. That water is a polar solvent, for instance, is as crucial to biochemistry as it is to atmospheric science and oceanography. For example, polar compounds dissolve typically in polar solvents: sugar is polar and successfully dissolves in the water in our bodies; for similar reasons, natural salts dissolved over the ages account for our saline ocean waters. Oil is not polar and fails to dissolve in water – as you will notice if you pour a spoonful of cooking oil in a cup of tap water.

Molecules distort as they vibrate and those molecules that happen to change their dipole moments when they vibrate can absorb infrared radiation and emit radiation back into the atmosphere, which contributes to atmospheric warming. Like BeF_2, carbon dioxide (CO_2) is a linear triatomic molecule with no dipole moment, but when it vibrates it bends and that changes the dipole moment. In the process, CO_2 absorbs and emits infrared radiation such that it is a so-called greenhouse gas and contributes to warming. Other non-polar molecules like methane (CH_4), which is tetrahedral, and sulfur hexafluoride (SF_6), which is octahedral (Figure AII.6), also distort when they vibrate and are considered to be greenhouse gases. Non-polar linear diatomic molecules like N_2 and O_2 only stretch when they vibrate and their dipole moments remain zero throughout, so they do not contribute to atmospheric warming. That's a rather fortunate situation for us since N_2 is the most abundant and O_2 is the second most abundant gas in the earth's atmosphere.

Figure AII.6 shows a few additional structures that are polar and others that are not. The polarities of the individual bonds is shown qualitatively along with the direction of the net dipole for the polar molecules. See if the directions of the resultant vectors (the net dipoles) make sense for you given the component dipole vectors.[3]

For three cases in Figure AII.6, we have indicated the presence of a lone pair with a dashed arrow. The lone pairs are often

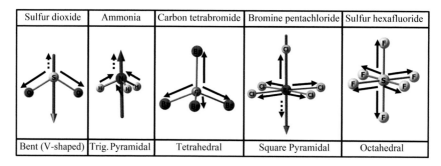

Sulfur dioxide	Ammonia	Carbon tetrabromide	Bromine pentachloride	Sulfur hexafluoride
Bent (V-shaped)	Trig. Pyramidal	Tetrahedral	Square Pyramidal	Octahedral

FIGURE AII.6 Qualitative representations of the bonds (full arrows), and lone pairs (dashed arrows) contributing to the net dipole (larger blue arrows) – not drawn to scale – of sample molecules. The terms used in chemistry to describe the shapes of the molecules are indicated below each structure. The trigonal pyramidal geometry is abbreviated 'Trig. Pyramidal' in the figure. The chemical formula of each molecule is, from left to right: $\underline{S}O_2$, $\underline{N}H_3$, $\underline{C}Br_4$, $\underline{Br}Cl_5$, and $\underline{S}F_6$. The central atom is underlined here in each case.

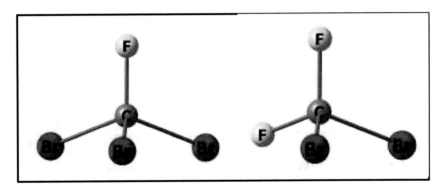

FIGURE AII.7   Model structures for CFBr$_3$ (left) and CF$_2$Br$_2$ (right).

not shown explicitly in molecular structures, but they can have a big effect on the size and direction of dipole moments. For CBr$_4$ and SF$_6$, the symmetry of each molecule is such that even though the individual bonds are polar, the overall molecule is not. Building a rough model of the molecules or thinking about the C and S centers in CBr$_4$ and SF$_6$ as the origin on a three-dimensional graph with four or six bond dipole vectors, respectively, will help you to see that the resultant vector is zero in both cases for the overall molecule. SO$_2$, NH$_3$, and BrCl$_5$ have less symmetrical electron distributions. The bond dipoles and the lone pairs do not cancel in those cases, so they are polar molecules. SO$_2$, NH$_3$, and BrCl$_5$ each happens to have one lone pair on the central atom, but a lone pair does not have to be present for a molecule to be polar. And there are molecules with lone pairs that are non-polar: XeF$_2$, for instance, is a linear molecule with three lone pairs arranged symmetrically on the central Xe atom and is non-polar.

Questions for you: Consider the CFBr$_3$ and CF$_2$Br$_2$ molecules shown in Figure AII.7. They both have tetrahedral-like geometries, with C as the central atom. Will (i) both, (ii) one, or (iii) neither of the two molecules be polar?

Note: There is definitely no lone pair on the central C atom. You only have to consider the geometry and the four bonds in each molecule. In thinking about your answer to this question, do not simply choose from the three options provided. Think, too, about how you would justify your answer in a short paragraph.

NOTES

1 Our dipole moment vectors run from positive to negative ($+ \rightarrow -$). The reverse direction is used by some for dipole moment vectors (there are old disagreements on this), but the magnitude of the net dipole moment is independent of that choice, as long as one sticks to their chosen convention.

2 Wharton, L.; Berg, R. A.; Klemperer, W. Geometry of the Alkaline-Earth Dihalides *J. Chem. Phys.* **1963**, *39*, 2023–2031.

3 It can be difficult to recognize the three-dimensional arrangements of the atoms in two-dimensions on the page. The arrangements are as follows: SO_2 is bent or V shaped with a lone pair (not shown) pointing up; NH_3 is pyramidal, with the N and its lone pair shifted out of the plane of the three terminal H atoms; CBr_4 is tetrahedral (like methane, CH_4), with the C sitting inside a tetrahedron with Br atoms at each vertex; $BrCl_5$ is a square pyramid – the Br is in the middle of a square base, with the fifth Cl atom above Br. A lone pair (not shown) pointing down, away from that axial Cl atom, has an impact on the size and direction of the $BrCl_5$ dipole moment. SF_6 is octahedral. The F atoms point to the vertices of an octahedron.

Appendix III

Safe Problem-Solving

Minimizing Risk while Solving Problems

Safe operations in teaching and research laboratories is critical. Professional organizations, laboratory managers, college faculty, and other stakeholders across the sciences agree on the importance of safe laboratory practices and working conditions, and we work to provide practical and effective guidelines in support of that goal.[1] People have suffered significant injury and even died as a result of attempting to answer scientific questions experimentally without the requisite training or due precautions.[2,3] Where the blame lies for those regrettable outcomes is not our question here – resolving that question can be difficult; but it should be emphasized that your actions can create unsafe environments for both yourself and others who share the laboratory space and resources with you. Everyone working in a laboratory or research area shares in the responsibility to make it as safe as possible. If you see a potential source of harm, you should respond – speak up, alert your neighbors, talk with your laboratory teaching assistant or professor. Better to be safe now than sorry later.

Some basic guidelines that you will receive in the typical teaching or undergraduate experimental research laboratory are included here. Guidelines may be tailored to specific environments (areas with biological hazards [viruses, bacteria, bio-waste of any sort, etc.], and radioactive, corrosive, or potentially explosive materials, for example), but the general ethos will be the same. If you are in doubt, always check with the safety or laboratory supervisor or lean toward the more conservative general guidelines of which you are aware.

- Review carefully any set of guidelines available for people working in your laboratory and any specialized direction for the particular experiment that you are working on at any given point.

- Keep safety at the front of your mind each step of the way as you plan your laboratory activities for the day.
- Dress such that there are appropriate barriers between your skin and chemicals. You will learn more in college about proper personal protective equipment.
- Wear appropriate eye protection – laboratory specific eyewear or attachments to regular glasses that provide more complete protection for the eyes.
- Wear closed-toe shoes that protect the back of your foot as well. If a spill occurs, you do not want to step in a caustic puddle in slippers, or to have chemicals dripping from the edge of the workbench onto your toes.
- Wear a lab coat if advised to do so. Whether you wear a lab coat or not, however, long pants are preferred. Avoid crop tops, shorts, ripped jeans, or short skirts, loose clothing, bulky jackets, long lanyards extending from your clothing, and any other item on your person that might get caught in or dislodge equipment, which can lead to injury or accidents as you move around the laboratory. For similar reasons, wearing long hair in a bun, under a hat, or tied back in some way, is good practice.
- Very close-fitting clothes can be harder to remove in the rare event that chemicals get onto your clothes while you are working but lab coats obviate that concern.
- Let your lab mates or lab partners know that you welcome their suggestions or requests if you seem to be doing something unsafe in the laboratory. You should feel free to let them know, too, if they seem to be operating unsafely. Establishing a generally positive rapport with other lab members early can help to create an atmosphere for having such conversations later on.
- Be aware of general and emergency exits.
- Be aware of the location of emergency equipment – fire extinguishers, eye wash stations, first aid kits, relevant shutoff valves, and showers. (Some laboratories have showers built in. They allow for a rapid response if chemicals get on to significant portions of your body such that washing at a faucet or sink is not enough.)
- Bags and other objects that you bring to the laboratory should be stored well out of your way in case you have to leave the room quickly. You do not want to trip over a bag as you turn to respond to an alarm or a fire.

- Do not play in the laboratory. Scientists just want to have fun, but the fun has to be limited in the laboratory to the excitement of a successful experiment done well and safely in a pleasant, supportive, and engaging environment.
- Arrive to the laboratory early or plan on staying a bit later if needed. Rush is the enemy of safety. Lack of sleep can have the same negative effect.
- Be sure that your chemical containers are labelled clearly and accurately. Chemical bottles from which you will be transferring chemicals – and these are usually well labelled by manufacturers – and the beakers, conical flasks, and other glassware into which you will transfer chemicals for dissolution or reaction should be labelled properly.
- Waste containers should be labelled clearly as well, and they should be stored properly when not in immediate use. The disposal of biological, chemical, and other types of waste is usually arranged ultimately by professionals at your institution. Your cooperation is usually required in the laboratory to ensure that different types of waste are disposed of in their specified containers.
- Do not point filled syringes, pipettes, or the narrow openings of reaction vessels that are in use toward others working around you. If you are directed to do so (for the entire laboratory session, or for a specific part of an experiment), be sure to work within designated enclosures or fume hoods to ensure the safety of everyone involved.
- Put laboratory equipment, especially bottles of chemicals, glassware, radioactive materials, and so on, away appropriately when you are finished with them.
- Clean up your area at the end of the class or after a day of research activities.

NOTES

1   See, for example: Howson, B. RAMPing up Safety Education: The Time Is Now *Chem. Eng. News.* **2016**, *94(18),* 35.
2   Kemsley, J. 10 Years Later, Where Are We? Chemists Discuss Their Efforts *Chem. Eng. News.* **2019**, *97(1)*, 16–23. [Ten years after Sheharbano (Sheri) Sangji's death from an accident in a chemistry laboratory at The University of California, Los Angeles, various stakeholders reflect on safety in the field. Contributors to sections in the

article were: Chemjobber, D. Decker, I. A. Tonks, P. J. Alaimo and J. Langenhan, N. Sampson and J. Rudick, and H. Thorpe.]

3 Lemonick, S. 25 Years after Karen Wetterhahn Died of Dimethylmercury Poisoning, Her Influence Persists (Chemist Left Legacies in Lab Safety, the Scientific Method, and Women in Science) *Chem. Eng. News.* **2022**, *100(21)*, 28–34. A version of this article also appears in: *ACS Chem. Health Saf.* **2022**, *29*, 327–332.

Index

Note: Page locators in *italics* indicate a figure.

A

abstract writing, 89–90
academic
    activities, 84
    assignment, 105
    commitment, 103, 116
    discipline, 1–2, 4, 7, 69–71
    excellence, 4
    growth, 104
    outcomes, 102
    research, 21
    skills, 5
    success, 102
accommodation, 3, 85, 95
accomplishment, 51, 102
action plan, 5, 123
active participation, 9, 21, 77, 79, 103
actively study, 10
ad hoc, 71
advisor, 3, 93–94
algebra, 45
    linear, 67, 114
    matrix, 66–67
analog clock, 85n5
analogy, 36, 46, 55, 104
analysis
    data, 61, 75, 77, 80
    error, 81
    logical, 8, 13, 22, 27, 36
    scientific, 57
anxiety, 6, 11–12, 35, 96, 101
appreciation, 3, 10, 39, 70, 81
approach
    alternative, 2, 49, 58, 67
    personal, 81, 93, 99
arena, 96, 106
argumentation, 3, 88
arguments, 34–35, 88
arrangements

atomic, *8,* 17n4, 29
    three dot, 37n1
aspiring, 71, 106
assemble, 10
assign (value), 20
assignment
    engagement, 84, 88, 105
    preparation, 76, 79, 88
    short-cuts, avoid, 24
assume, 6, 25, 37, 49, 88
assumption, 90, 101–102
atoms
    formations, 8, 24, 29–30
    identical, 117
    reactive, 20, 22
attractive (challenges), 2
audience, 84, 96–97

B

best option, 17
best-fit, 61
bias, 69, 76
bravery, 2
brevity, 33, 90–92
building
    blocks, 17n4, *see also* atoms
    rapport, 78
    as structure, 52, 71

C

calculation, 24, 29, 77, 83, *see also* solution
calculus, 45, 69
captions, 88, 93
cause, 21, 36
challenges
    problem-solving, 1–2, 5, 79
    taking on new, 81, 94, 102, 105

check
   guess and check method, 2, 16, 17n1
   logic, 28, 39
   solution, 60
   verify (as with instructor), 35, 43, 77, 83, 88
chemical
   allergy, 84–85
   formula, 8, 29–30
   handling, 77
   reaction, 19–20, 22, 31n2, 72n6
chemistry
   course, 30, 45, 70
   student, 7, 16, 29
choice (personal), 6, 28, 91, 97, 104
circle, 67–68, 70, 71n1
Civil War, 24
clarify, 39, 48, 97
clarity, 90–92, 96
classroom, 1, 3
   discussions, 87, 106
   experience, 37, 79–80, 85
coach, 42
college
   campus, 4, 94, 97, 104
   community, 4, 85, 104
   courses, 5, 10, 69, 80, 82, 91, 106
   experience, 4–5, 81, 104
   faculty, 4, 70, 85, 94, 101
   students, 34, 102, 106
commitment
   faculty, 3
   overcommitment, 103
   student, 4
common
   mathematical analysis, 61–62
   notions, 20
commonly, 36, 45, 48
communication, 43n3, 87
computers, 4, 61, 67, 70–71
concentration, 15, 25, 28–29, 58
confidence, 2, 6, 12, 35, 46, 69, 96
constant
   exponent, 57–58
   law of mass, 19
   multiple, 59, 60
   of proportionality, 67
   unitless, 27
control, 78, 82, 96
converge, 24
Cornell note-taking method, 11, 17n5
Cosine rule, 51
count your blessings, 12, 35
courses
   college, 5, 10, 67, 71, 91
   science, 19, 70

style, 71, 78–79
create, 19, 84
creative, 8, 42, 71
creativity, 88
critical thinking, 8, 50, 75, 106
cuboids, 68
cylinder, 67–68, 82

D

data
   experimental, 61, 75, 82
   mining, 95
   points, 55, 61–65
   processing, 15
debate, 4, 70, 105
deceive, 84
decisions, 90
defects, 82
delude, 84
design, 71, 75–76, 79, 91
development (growth), 3, 69, 83, 94, 104
diagonal, 66
diagram, 12, 42, 52, 93, 98, *see also* Venn
         Diagram
diameter, 67–68
differences, 90
different
   approaches, 28–29, 78, 84
   arrangements, *8*, 17n4, 29–30
   functions, 61, 64
   processes/procedures, 42, 52, 88
   properties, 12, 17n4, 35
difficulties, 28, 84, 90
disagreement, 85n1
discipline
   academic, 7–8, 20, 71n1
   personal, 2, 4, 87
   scientific, 45, 55, 70
diverse, 78, 99
diversion, 51, 99
diversity, 98
doctors, 50
document, 77, 89
dreams, 76

E

economics, 55, 71
effect, 35–36, 107, *see also* Mpemba effect
effective, 5, 7, 11, 78, 87–88
engagement, 1–3, 21, 39, 41, 84, 90, 102
entrepreneur, 71
equation
   linear (straight-line), 56–58, 62

quadratic, 14, 58–60
    simultaneous, 62–65, 67, 109–110
errors, appreciation of, 77–78, 81
essay writing, 37, 88–90
ethics, 3, 84, 87
Euclid's Elements, 20–21
evidence, 21, 75–76, 83, 102
exam, 5, 11, 24, 42, 87, 105, *see also*
       test taking
example
    equation, 57–58
    experimental data, 61
    simultaneous equation, 64–65
expectation, 75
experience
    educational, 3–5, 62, 69, 101–102, 104
    learning, 77, 79, 95–96, 99
experiments matter, 75, 79
explanation, 21, 35–37, *see also*
       observation

F

fact, statement of, 80
factor, 22–23
factual, 96
faculty, 3–4, 70, 85, 94, 101, *see also* instructor;
       professor
fail, 5, 13, 88
failure, 75
falsify, 84
farmers, 50
feedback, 3, 77, 89, 95, 99
focus, 2, 11, 69, 77, 106
footnotes, 88
formula, 2, 27–28, *see also* chemical; quadratic
       formula
functions
    exponential, 54–55, 64
fundamental, 24–25, 50
funding, 93
future
    direction, 91, 98
    learning, 2–3, 77

G

goals
    personal, 5, 16, 88, 104
    professional, 3
graphs, 55, 60–61, 78, 92–93
group, 94, 96–97, *see also* study group
grow, 45, 69, 95, 104
growth, 3–4, 69, 104
guess, 2, 16n1, 80

H

heart, 2, 79
helpful
    definitions, 54
    feedback, 88–89
    suggestions, 36, 42, 49, 57
hints, 10, 91
hopes, 76
hypothesis, 76, 85n3, 96

I

ideas, 20–21
    expressing, 35–36, 41–42
    logical, 22
    mathematical, 47, 71
impact, 60, 82, 88, 90
Implicit Faculty commitment, 3
importance of self-care, 4
important (steps to problem-solving), 9, 45, 90, 95
impostor syndrome, 102
improve, 77, 95
improvement, 77
independence (intellectual), 3, 49
independently thinking, 9, 11, 71, 81
influence, 60, 82, 89, 96, 99
innovation, 88
instruction, 3, 35, 36, 70, 76
instructor
    guidance, 1, 10, 71
    relationships, 81–84
instruments, 82, 85n5
integrity, 3, 84
intellectual
    engagement, 3, 90, 102
    independence, 3, 69, 104
    thinking, 50, 106
intention, 4, 69
interaction, 3, 78
intercept, 57–58, *59*, 60
interesting relationships, 48, 53
international system of units, 24, 31n8
internship, 105
intervention, 105
intriguing, 4, 54–55, 60, 77
introduce, 66–67, 95
introduction
    abstract, 91, 93, 97
    mathematical, 67
introductory
    slides, 97
    statistics, 62
    summary, 89
isomers, 8, 17n4

J

job
market, 70, 104
your job, 4, 13
joy, 2, 7, 41, 77, 79, 99

K

knowing, 7–8
knowledge
mathematical, 45, 69
personal, 7–8, 15, 61, 81–82, 96, 101
scientific, 30, 31n5, 75–76, 87

L

laboratory
equipment, 79, 82
experience, 79–80, 84
experiments, 24, 75–77, 80
instructor, 82, 84
lawyers, 50
layers, 69–70
learn
how to, 7, 9, 17n2, 34, 99, 102
need to, 39, 67, 71, 95
learning
active, 77, 79
experience, 3, 77
language, 9, 35
rote, 27
strategies, 11–12, 36
librarians, 89
limiting factors, 22
Lincoln, Abraham, 24
linear
algebra, 67
combination, 113–114
equations, 56
form, 58, *61*, 63–64
molecule, 121
liquid, 40, 85n5
logarithms, 55–56
logic
of amounts, 23
general scientific, 22–23
problem solving, 27–28, 31n9
subject-independent, 19, 70–71
symbolic, 35–36

M

manage (self), 77, 79, 96
management (time), 79
manufacturing, 22–23

mass conservation, law of, 19–20
master/mastery, 7, 10, 39, 42, 101–102
material
learning, 10–11, 39, 70, 88, 101
reading, 35, 40, 95
roofing, 52
mathematics
basic, 45
results, 28
strategies, 2, 50, 67
training, 70–71
matrices, 65–67, 69
matrix
algebra, 67
square, 66
maximum, 14, 68
measure, 25, 29
measurement, 77, 82
memorization, 2, 5, 9, 27
mentor, 3, 81, 93–94, 97
mentoring, 69
mind
habits of, 3
suggestions, 9, 19, 35, 83, 89, 90, 101
mistakes, 77, 84
mitigation, 12
molecules, 24, *29*, 30
more effective, 5
more time, 5
Mpemba effect, 40

N

negative
equations, 57, 59
influence, 82
number, 23–24
perception, 69
Newton, Isaac, 21

O

object, 23, 68
objective, 4, 10, 51, 66, 104
omniscient, 96
opportunity, 1
option, 2, 10–11
selection, 63–64
oral presentation, 87, 95
organize, 10, 13, 75, 89, 97
outcome
academic, 77, 102
desired, 5, 84, 102
key objective, 91, 93
oversight, 96

P

partnership, 3, 79
patience, 6, 81
peer support, 4, 9, 71, 89, 94, 102–103
perimeter, 67–68, 70–71
persistence, 4, 6, 69
persisting, 101
plagiarize, 84, 88–89
plan, 5, 97–98, 103
position, 14, 87, 111–112
possibilities (unknown), 83
posters (preparing), 91
powers, 56
practice
    problems, 40, 42
    questions, 9, 42, 65, 81
preparation, effective, 5, 7, 42, 80, 95–96
preparedness, 2, 9, 101
presentation, 3, 39, 95–99
prisms, 68
problem
    blessings of, 12
    defined as, 1
    logic, 19
    preparation, 5
    preparation for, 5
    propose, 3
    solver/solving, 4, 6–7, 9, 14
    study strategies, 9
    taking on, 9
    test taking, 11
problem-solving
    intellectual, 1
    logic, 22
    mentor, 81
    skills, 48
    steps, 6–7, 28, 103–106
professor
    academic ability, 7, 27
    rapport building, 10–11, 35, 81, 101
    responsibilities of, 2–4, 69, 84m 88–89
properties, 54–55, 58
proportionality, 67
proposition (problem), 2, 8
Pythagoras' theorem, 51

Q

quadratic
    curve, 59–60
    equation, 14, 58–60
    formula, 2, 13, 60
    function, 60
questions
    practice, 42, 65

research, 94
    scientific, 76

R

radioactive, 57–58
radius, 67–68, 70, 71n1
reaction, 15
    chemical, 16, 19–20, 22, 31n2
reading skills, 2–3, 9, 29, 40, 42, 79, 88
reason, 84, 90, 101, 105
reasonable, 13, 23–24, 84, 88, 96
recognize, 16, 57
rectangles, 68
reminder, 45, 77
requirement, 87
research
    academic, 21, 76, 84, 91, 93, 98–99
    opportunities, 1, 69, 95, 105
    publications, 89
responses, 2, 35, 37
results
    experimental, 83
    key, 91, 95, 98
    scientific, 47, 77, 82
reverse, 49

S

scientific
    laws, 21
    methods, 21, 75–76, 85n1
    report, 89–91
    results, 47, 77, 82
    rules, 21
service, 85, 104, 106
shapes, 67, 70
simple
    equations, 28
    functions, 67
    molecule, 29, 43
    steps, 2, 16
simplify, 22, 50, 55
Sine Rule, 50
skills, 35, 43, 50, 69
    academic, 5, 10, 35, 69, 102
    communication, 87
    mathematical, 46, 79
solution
    basic math, 45
    colorless, 80
    counting blessings, 49–50
    logical, 28–29
    possible, 2, 7, 12–13, 25, 81
    practical, 75
sphere, 68

square, 70
    matrix, 66–67
standards, 3, 80, 84
step (simple), 2, 16
stoichiometric, 16, 77
student
    academic responsibility, 4–5, 11–12, 27
    commitment, 4, 102
    commitment to, 3, 8, 10
    learning styles, 16, 23, 26
    requirements, 76
    undergraduate, 6, 21, 53, 69, 97, 104
study group, 11, 43, 78, 91
style
    conversational, 91–92, 99
    essay, 37, 88–89
subtraction, 46
success
    academic, 2, 4–5, 13–14, 79
    willingness for, 81, 87, 93
summary (abstract), 89–90, 92–93
support
    academic skills, 9, 11, 35, 69, 93, 102
    peer, 4, 71, 102
syllabus, 3, 10
symbols, 33–35

T

talks (preparing and giving), 95
teaching assistant (T.A.), 77–78, 85
team work, 11, 77, 95
tendency, 12
test taking, 1, 9, 11–12, 81
textbooks
    benefits from, 41–42
    as coach/mentor, 42–43
    usefulness of, 39–40
thermodynamics, 19

thwart (efforts), 1
titration, 29, 31n10
topics (key), 10–12, 70, 94, 96, 102
trial and error, 2, 16n1, 17n6
trial and error strategy, 2, 16n1
triangles, 47–48, 50–51, 68
trigonometric ideas, 47
trust building, 2–3

U

unavoidable, 82
undergraduate
    curriculum, 4, 45, 53, 70
    lab work, 75–76, 81
    poster program, 91, 94
unintentional, 82
units
    basics of, 24
    meaning and, 25

V

vectors, 69
Venn Diagram, 34
volume, 25, 28–29
    measuring, 68, 77, 82

W

words
    short answers, 35–37
    word problems, 33–34
writing skills, 37, *see also* essay

Z

zeroth law, 19–20